THE 2011 NATIONAL ELECTRICAL CODE
BOOK OF IN-DEPTH CALCULATIONS

VOLUME 3

Covering selected Articles
of Chapters 4 through 7
and the Tables of Chapter 9

Alvin J. Walker

THE 2011 NATIONAL ELECTRICAL CODE BOOK
OF IN-DEPTH CALCULATIONS

VOLUME 3

© 2015 Alvin J. Walker

Portions of this material are reproduced from NFPA70®- 2011, *National Electrical Code*®,
Copyright © 2010, National Fire Protection Association, Quincy, MA.
This reprinted material is not the complete and official position of the NFPA on the referenced
subject, which is represented only by the standard in its entirety. The calculation accuracy
of any equation is the sole responsibility of the author and not the NFPA.

NFPA70®, *National Electrical Code* and *NEC*® are registered trademarks of the
National Fire Protection Association, Quincy, MA.

ISBN 13: 978-0-9831358-4-5

LCCN 2015907070
First Edition
1 2 3 4 5 6 7 8 9 10

Walker & Walker Electrical Consultants
For more information, please contact
Alvin Walker
318-393-6841
www.alvinwalker.com

TABEL OF CONTENTS

INTRODUCTION TO VOLUME 3 . . . .**xv**

Number in brackets indicates the number of questions per NEC sections. Total questions: 244

CHAPTER 4 - Equipment for General Use

ARTICLE 400
Flexible Cords and Cables

ARTICLE 408
Switchboards and Panelboards **2**

ARTICLE 409
Industrial Control Panels **2**

ARTICLE 422
Appliances

ARTICLE 424
Fixed Electric Space-Heating Equipment

CHAPTER 5 - Special Occupancies

ARTICLE 517
Health Care Facilities

ARTICLE 520
Theaters, Audience Areas of Motion Picture and Television Studios, Performance Areas, and Similar Locations

ARTICLE 530
Motion Picture and Television Studios and Similar Locations

ARTICLE 550
Mobile Homes, Manufactured Homes, and Mobile Home Parks

ARTICLE 551
Recreational Vehicles and Recreational Vehicle Parks

ARTICLE 552
Park Trailers

ARTICLE 692
Fuel Cell Systems

ARTICLE 695
Fire Pumps

CHAPTER 7 - Special Conditions

ARTICLE 725
Class 1, Class 2, and Class 3 Remote-Control, Signaling, and Power-Limited Circuits

CHAPTER 9 - Tables

CHAPTER 9
Tables

INTRODUCTION TO VOLUME 3

As it would seemingly follow, Volume 3 is an extension of Volume 2, only more intriguing. Covering Chapters 4 through 7 and Chapter 9 of the National Electrical Code (NEC), Volume 3 contains Articles 430 and 450 - elite articles - in which the user will find a wealth of detailed information and a sound means to obtaining a better knowledge of all related calculation requirements.

Also in this volume, the user will find a vast arrangement of additional articles often not referenced, or even unknown, where electrical calculations are required.

Volume 3 provides a full comprehension of the Tables of Chapter 9, as they relate to the use and applications of sizing raceway, as well as the performing of practical voltage drop calculations.

In considering the overall layout of Volume 3, the following color scheme was used to provide a consistent path for quick reference of selected material where:

- red identifies those articles of the NEC involving electrical calculations.
- the color green is used to identify each specific section of an article that requires some type of an electrical calculation.
- followed by the color blue which is used to identify questions that are relative to each given section.
- Answers and explanations in response to each question are identified in black.
- Where there is a cross-reference of related questions deep-purple is used.
- Finally, the color brown is used to identify NEC material, supplements, discussions and pertinent information.
- Also of important note, where sections contain multiple questions, the relative question numbers are enclosed in parenthesis ( ), followed by the total number of questions in brackets [ ]. Volume 2 also features large-print for easier visibility and reading.

Combined, Volume 3 contains over 240 questions and answers.

ARTICLE 400 - Flexible Cords and Cables

Table 400.5(A)(1) - Allowable Ampacity for Flexible Cords and Cables *and*
Table 400.5(A)(3) - Adjustment Factors for. . . Flexible Cord or Cable

1. A 12/4 Type HPD flexible cord is used for an installation that requires all enclosed conductors to be current-carrying. What is the allowable ampacity of the cord's enclosed conductors?

Table 400.5(A)(1) lists an allowable ampacity of 30 amps for a 12 AWG conductor with type HPD insulation. Because the four conductors within the flexible cord are current-carrying, the ampacity of the conductors must be reduced. Table 400.5(A)(3) provides adjustment percent values to reduce the current when more than three current-carrying conductors are present. In this case the adjustment percent value is .80, therefore 30 amps x .8 = 24 amps.

Based on the results, the allowed ampacity of the conductors enclosed within the cord is 24 amps.

Table 400.5(A)(2)) - Ampacity of Cable Types

2. A 10/3 SCT cable with ground rated for 194°F is routed through an area where the ambient temperature is expected to reach 105°F. Is the cable permitted to operate at its allowable ampacity?

According to NEC 400.5, where cords are used in ambient temperatures exceeding 30°C(86°F), the temperature correction factors from Table 310.15(B)(2)(a) that correspond to the temperature rating of the cord shall be applied to the ampacity from Table 400.5(A)(2).

Per Table 400.5(A)(2) [see footnote 3], the allowable ampacity of a No. 10/3 SCT cable rated for 194°F is 49 amperes. Based on the correction factors of Table 310.15(B)(2)(a), the applicable correction factor for a conductor being exposed to an ambient temperature of 105°F is .87. When applied, the adjusted ampacity of the cable's conductors is, 42.63A (49A x .87).

ARTICLE 408 - Switchboards and Panelboards

Panelboard *versus* Load Center

Quite often the question is asked, "What's the difference between a load center and a panelboard?" In abbreviated terms according to Article 100 a **panelboard** is actually the interior components or simply "the guts" which are designed to be *field- assembled* in a cabinet or cutout box and are accessible only from the front. Although not recognized or defined by the NEC, a **load center** unlike a panelboard is *factory-assembled* with the interior components mounted in an enclosure. A load center must meet the same requirements per Article 408 as a panelboard.

After the 2005 Edition of the National Electrical Code, NEC 408.34(A) (Lighting and Appliance Branch Circuit Panelboards), 408.34(B) (Power Panelboards) and 408.36(B) (Power Panelboard Protection) were deleted thereby leaving Article 408 without the need for any calculations.

ARTICLE 409 – Industrial Control Panels

409.21(C) - Rating (Overcurrent Protection)

See question pertaining to 670.4(C) which has calculations similar to the requirements of this section.

ARTICLE 422 - Appliances

422.11(E)(2) - Single Non-motor-Operated Appliance [rated 13.3 amps or less] (Overcurrent Protection)

1. If unmarked on the appliance, what size overcurrent device is required to protect a 125V, 1600W commercial toaster? What size branch-circuit conductors are required?

DETERMINE APPLIANCE CURRENT (I) RATING

$$I = 1600W / 125V = 12.8A$$

If the rating of the overcurrent device is not marked on the appliance and the appliance is rated less than 13.3 amperes, the overcurrent device cannot exceed 20 amperes according to NEC 422.11(E)(2).

As a result, the single commercial appliance can be protected by an individual overcurrent device rated for 15A as a minimum or at a maximum of 20A. See NEC 210.23 (Volume 2) and 422.10(A).

BRANCH-CIRCUIT CONDUCTORS

If *copper*, the branch-circuit conductors supplying the commercial toaster must be sized according to **NEC 240.4(D)(3)** [14 or 12 AWG for 15A device] or **NEC 240.4(D)(5)** [12 AWG for 20A device].

If (hypothetically) *aluminum*, the branch-circuit conductors supplying the commercial toaster must be sized according to **NEC 240.4(D)(4)** [12 AWG for 15A device] or **NEC 240.4(D)(6)** [10 AWG for 20A device].

422.11(E)(3) - Single Non-motor-Operated Appliance (rated over 13.3 amps) (Overcurrent Protection) (2. - 3.) [2]

2. What size overcurrent device is required to protect a single-phase, 240-volt, 4500W electric water heater when the rating of the overcurrent device is not marked on the nameplate of the appliance? What size overcurrent device is required if rated for 208-volts?

DETERMINE APPLIANCE CURRENT (I) RATING

$$I = 4500W / 240V = 18.75A$$

If the rating of the overcurrent device is not marked on the appliance and the appliance is rated over 13.3 amperes, the overcurrent device cannot exceed 150 percent (1.50) of its current rating according to NEC 422.11(E)(3). As a result,

$$18.75A \times 1.50 = 28.13A$$

Because NEC 422.11(E)(3) permits the next higher standard rating overcurrent device, a 30A overcurrent device can be used.

DETERMINE APPLIANCE CURRENT (I) RATING

$$I = 4500W / 208V = 21.63A$$

Again, if the rating of the overcurrent device is not marked on the appliance and the appliance is rated over 13.3 amps, the overcurrent device cannot exceed 150 percent (1.50) of its current rating according to NEC 422.11(E)(3). As a result,

$$21.63A \times 1.50 = 32.45A$$

Because NEC 422.11(E)(3) permits the next higher standard rating overcurrent device, a 35A overcurrent device can be used.

3. Assuming the water heater in question No 4. was not marked with the rating of an overcurrent protective device, determine the size overcurrent device needed to protect the water heater.

In accordance with NEC 422.11(E)(3), based on the rated current (68.53A) of the water heater per calculation yields,

$$68.53A \times 1.50 = 102.8A$$

Because NEC 422.11(E)(3) permits the next higher standard rating overcurrent device, a 110A overcurrent device can be used.

422.13 - Storage-Type Water Heaters (4. - 5.) [2]

4. What size THW copper branch-circuit conductors are required to supply a 230V, 27.3kW, 3φ water heater that's rated for a maximum capacity of 100 gallons?

According to NEC 422.13, a fixed storage-type water heater that has a capacity of 120 gallons or less *shall be considered a continuous load** for the purposes of sizing branch circuits. Therefore, the conductors are required to be sized no less than 125 percent of the water heater's nameplate rating. Based on the information provided, the ampere rating (I) of the water heater is calculated where,

$$I = \frac{27,300W}{230V \times 1.732} = 68.53A \times 1.25 = 85.66A$$

*Means, regardless of whether the load will operate continuous [for three hours or more] treat it as if it would, size the branch-circuit conductors accordingly.

As a minimum, 3 AWG THW copper conductors rated for 100A can be used per Table 310.15(B)(16).

5. At 75°C, what size copper conductors are required to supply the water heaters in question No. 2.? What size conductors are required, if supplied by Nonmetallic Sheathed Cable (Romex)?

In accordance with NEC 422.13, the calculated current ratings of both water heaters must be increased by 125 percent to size the branch-circuit conductors per rated voltage (water heaters considered to be continuous loads).

At 240V - 18.75A x 1.25 = 23.44A   **At 208V** - 21.63A x 1.25 = 27.04A

At **240V**, based upon 23.44A, per Table 310.15(B)(16), 12 AWG copper which has a rated ampacity of 25A at 75°C can be used as the branch-circuit conductors. However, in accordance with NEC 240.4(D)(7), the conductors must be sized no less than 10 AWG based on the use of a 30A overcurrent device.

At **208V**, based upon 27.04A, per Table 310.15(B)(16) and the use of a 35A overcurrent device, 10 AWG copper which has a rated ampacity of 35A at 75°C can be used as the branch-circuit conductors.

In reference to the use of Nonmetallic Sheathed Cable, per NEC 334.80, the allowable ampacity of the cable shall not exceed that of a 60°C (140°F) rated conductor.

Therefore, based upon the use of a **30A** overcurrent device per NEC 240.4(D)(7), 10/2 AWG Nonmetallic Sheathed Cable which is rated for 30A at 60°C must be used per Table 310.15(B)(16).

Based upon the use of a **35A** overcurrent device, 8/2 AWG Nonmetallic Sheathed Cable which is rated for 40A at 60°C must be used per Table 310.15(B)(16). Although NEC 240.4(B) would permit a 35A overcurrent device to protect 10/2 AWG Nonmetallic Sheathed Cable which is rated for 30A at 60°C, NEC 240.4(D)(7) only permit 10 AWG copper to be protected by an overcurrent device rated no higher than 30A.

ARTICLE 424 - Fixed Electric Space-Heating Equipment

424.3(B) - Branch Circuit Sizing (Branch Circuits) (1. - 2.) [2]

1. Three 250V-1800W electric baseboard heaters are being installed in an area where additional heating is needed. Determine the minimum branch-circuit conductor's ampacity and overcurrent protection device required for this installation.

DETERMINE LOAD CURRENT (I) RATING

$$I = 1800W \times 3 / 250V = 21.6A$$

NEC 424.3(B) states, *fixed electric space-heating equipment shall be considered a continuous* load*. Considering such requirements, an electric baseboard heater which is categorized as fixed electric space-heating equipment must have branch-circuit components sized as so to serve the baseboard heaters. Referring to the general requirements of Article 210 for branch circuits, NEC 210.19(A) and 210.20 respectively requires the ampacity of branch-circuit conductors and overcurrent protection to be sized at 125 percent when supplying loads that are continuous.

*Means, regardless of whether the load will operate continuous [for three hours or more] treat it as if it would, size the branch-circuit conductors accordingly.

Therefore, applying the calculated load of the heaters and the 125 percent increase,

$$21.6A \times 1.25 = 27A$$

Based on the calculated results, the branch-circuit conductors are required to have a minimum ampacity of 27 amps where at minimum, size 10 AWG copper conductors are required. The overcurrent device is permitted to be sized according to NEC 240.4(D)(7) which requires the device to be sized at 30 amps, the largest size permitted per NEC 424.3(A).

2. A 9.5kW electric furnace with 75°C terminals is being installed in a dwelling unit. The furnace is equipped with a 2.7A blower. What size non-metallic sheathed cable is required to supply the furnace? What size overcurrent device is required to protect the furnace?

DETERMINE HEATER'S CURRENT (I) RATING

$$I = 9500W / 240V = 39.58A$$

Total Load

$$39.58A + 2.7A = 42.28A$$

Based on the explanation provided in question No. 1. per NEC 424.3(B) and the related sections of Article 210, the furnace and blower are both required to be sized no less than 125 percent of the total load resulting to,

$$42.28A \times 1.25 = 52.85A$$

Because the ampacity of non-metallic sheathed cable is limited to 60°C (NEC 334.80) the ampacity of the required cable is so referenced. Per Table 310.15(B)(16), 6 AWG non-metallic sheathed cable (where the ampacity at 60°C is rated for 55A) is required.

A 60 amps overcurrent device is required for overcurrent protection and is allowed per NEC 240.4(B).

424.22(B) - Resistance Elements (Overcurrent Protection)

3. What size overcurrent protection is required for a 240V single-phase 20kW electric space-heating unit containing resistance-type heating elements (coils)?

The utilization of resistance-type heating elements is the typical design for most electric space-heating units (furnace). Based on the given wattage rating of the heating unit at 240V single-phase the current rating of the unit is,

$$\frac{20,000W \ (20kW)}{240V} = 83.33A$$

Per NEC 424.22(B), resistance-type heating elements in electric space-heating equipment must be protected at not more than 60 amperes. When the heating elements in electric space-heating equipment are rated for more than 48 amperes, the elements are required to be subdivided and each subdivided load is limited to 48 amperes. Based on this requirement, the use of dual-circuits is implied which requires the heating elements of the electric-space heating equipment to be protected by dual overcurrent devices.

In this situation, the rating of the heating elements is 83.33 amperes which exceeds the 48 amperes limitation. As a result, the total load is required to be subdivided where the first 48 amperes of the element's rated current is considered followed by the remaining load current at 35.33 amperes (83.33A – 48A).

At 48 amperes, the overcurrent protection is determined based on the provisions of NEC 424.3(B) [continuous loads increased by 125 percent, also see section 210.20(A) for branch-circuit ratings] where,

$$48A \times 1.25 = 60A$$

As a result, a 60A overcurrent device is required based on the given calculation. The 60A overcurrent protection is the first device of the dual circuit required for this installation.

Now considering the remaining 35.33A and applying the same provisions of NEC 424.3(B) to determine the second required overcurrent device of the dual circuit,

$$35.33A \times 1.25 = 44.16A$$

As a code minimum, a 45A overcurrent device is required based on the given calculation. In general, for this installation a 60A and 45A overcurrent device is required for the protection of the subdivided heating elements of the electric-space heating equipment.

NEC 424.3(B) also applies for sizing the supplying branch-circuit conductors for this unit. Therefore, if this unit was used in a residential setting a 4/2 AWG non-metallic sheathed cables (at 70A/60°C) and a 6/2 AWG non-metallic sheathed cable (at 55A/60°C) would be minimally required per NEC. On the other hand, if this unit was used in a commercial setting 6 AWG (65A) and 8 AWG (50A) copper conductors at 75°C would be minimally required for the 60A and 45A circuits.

424.36 - Clearance of Wiring in Ceilings

4. Electric space heating cables are installed 2" above a heated ceiling. The cables will supply a 19 amps load. What size THW copper conductors are required to supply the heating cables?

The conductors supplying the heating cables must be rated no less than 125 percent of the rated load per NEC 424.3(B) because the scope (NEC 424.1) of Article 424 identifies heating cable in the same manner as any other type of heating equipment. Therefore,

$$19A \times 1.25 = 23.75A$$

NEC 424.36 states that wiring above heated ceilings must be considered as operating at an ambient temperature of 50°C (122°F) and the (wiring) ampacity requires derating based on the correction factors of the 0 - 2000 volt ampacity tables of Article 310.

Using the correction factors of Table 310.15(B)(2)(a) at 50°C, the ampacity of a THW copper conductor is required to be derated by 75 percent (.75). To determine the required size of the conductors, the following calculation is performed,

$$\frac{23.75A}{.75} = 31.67 \text{ amps}$$

where 10 AWG THW copper conductors can be used per Table 310.15(B)(16). However, if NEC 110.14(C)(1)(a) is applicable, 8 AWG THW copper conductors are required.

424.72(A) - Boiler Employing Resistance-Type Immersion Heating Elements in an ASME Rated and Stamped Vessel (Overcurrent Protection)

5. What size overcurrent protection is required to protect a 208V, three-phase, 125kW resistance - type boiler utilizing immersion heating elements?

Based on the given wattage rating at 208V, the three-phase current rating of the boiler is,

$$\frac{125,000W}{208V \times 1.732} \ (125kW) = 347A$$

Per NEC 424.72(A), a boiler employing resistance-type immersion heating elements contained in an ASME (American Society of Mechanical Engineers)-rated and stamped vessel shall have the heating elements protected at not more than 150 amperes. Where such boiler is rated more than 120 amperes, the heating elements must be subdivided into loads not exceeding 120 amperes.

Based on this requirement, the use of multiple-circuits is implied which requires the boiler to be protected by multiple three-pole overcurrent devices.

In this situation the rating of the boiler's heating elements is 347 amperes which exceeds the 120 amperes limitation. As a result, the total load is required to be subdivided into three parts where 120 amperes of the element's rated current is considered twice (347A – 120A – 120A) followed by the remaining load current at 107 amperes (347A – 240A).

At 120 amperes, the overcurrent protection is determined based on the provisions of NEC 424.3(B) [continuous loads increased by 125 percent also, see section 210.20(A) for branch-circuit ratings] where,

$$120A \times 1.25 = 150A$$

As a result, two 150A overcurrent devices are required based on the given calculation.

Now considering the remaining 107A and applying the same provisions of NEC 424.3(B) to determine the third required overcurrent device of the circuit,

$$107A \times 1.25 = 133.75A$$

For this installation, 3-three pole 150A overcurrent devices are required for the protection of the subdivided heating elements of the electric-space heating equipment.

424.72(B) - Boiler Employing Resistance-Type Heating Elements Rated More than 48 Amperes and Not Contained in an ASME Rated and Stamped Vessel (Overcurrent Protection)

See question pertaining to 424.22(B) and question Nos. 1. and 2. of section 424.3(B) which has calculations similar to the requirements of this section.

424.82 - Branch Circuit Requirements (Electrode-Type Boilers)

See questions pertaining to 424.3(B) which has calculations similar to the requirements of this section. For electrode-type boilers, motor loads are not included in the total load.

424.95(B) - Interior Walls (Wiring behind heating panels or heating panel sets)

6. Heating panel sets are installed in the interior walls of a building. If certain wiring behind the heating panel sets will supply a load of 17.8 amps, what size THHW (90°C) copper conductors are required to supply the load?

The same procedure is basically followed as with electric space heating cables.

$$17.8A \times 1.25 = 22.25A$$

NEC 424.95(B) states that wiring behind heating panels or heating panel sets located in interior walls or partitions must be considered as operating at an ambient temperature of 40°C (104°F) and the (wiring) ampacity requires derating based on the correction factors of the 0 - 2000 volt ampacity tables of Article 310.

Using the correction factors of Table 310.15(B)(2)(a) at 40°C, the ampacity of a THHW copper conductor at 90°C is required to be derated by 91 percent (.91).

$$\frac{22.25A}{.91} = 24.45 \text{ amps}$$

14 AWG THHW copper conductors can be used per Table 310.15(B)(16). However, if NEC110.14(C)(1)(a) is applicable, the ampacity of the 14 AWG THHW conductors is limited to 15 amps. Therefore, 10 AWG THHW copper conductors which have an ampacity of 40 amps but limited to 30 amps at 60°C are required.

ARTICLE 426 - Fixed Outdoor Electric Deicing and Snow-Melting Equipment

426.4 - Continuous Load

See questions pertaining to 424.3(B) which has calculations similar to the requirements of this section.

ARTICLE 427 - Fixed Electric Heating Equipment for Pipelines and Vessels

427.4 - Continuous Load

See questions pertaining to 424.3(B) which has calculations similar to the requirements of this section.

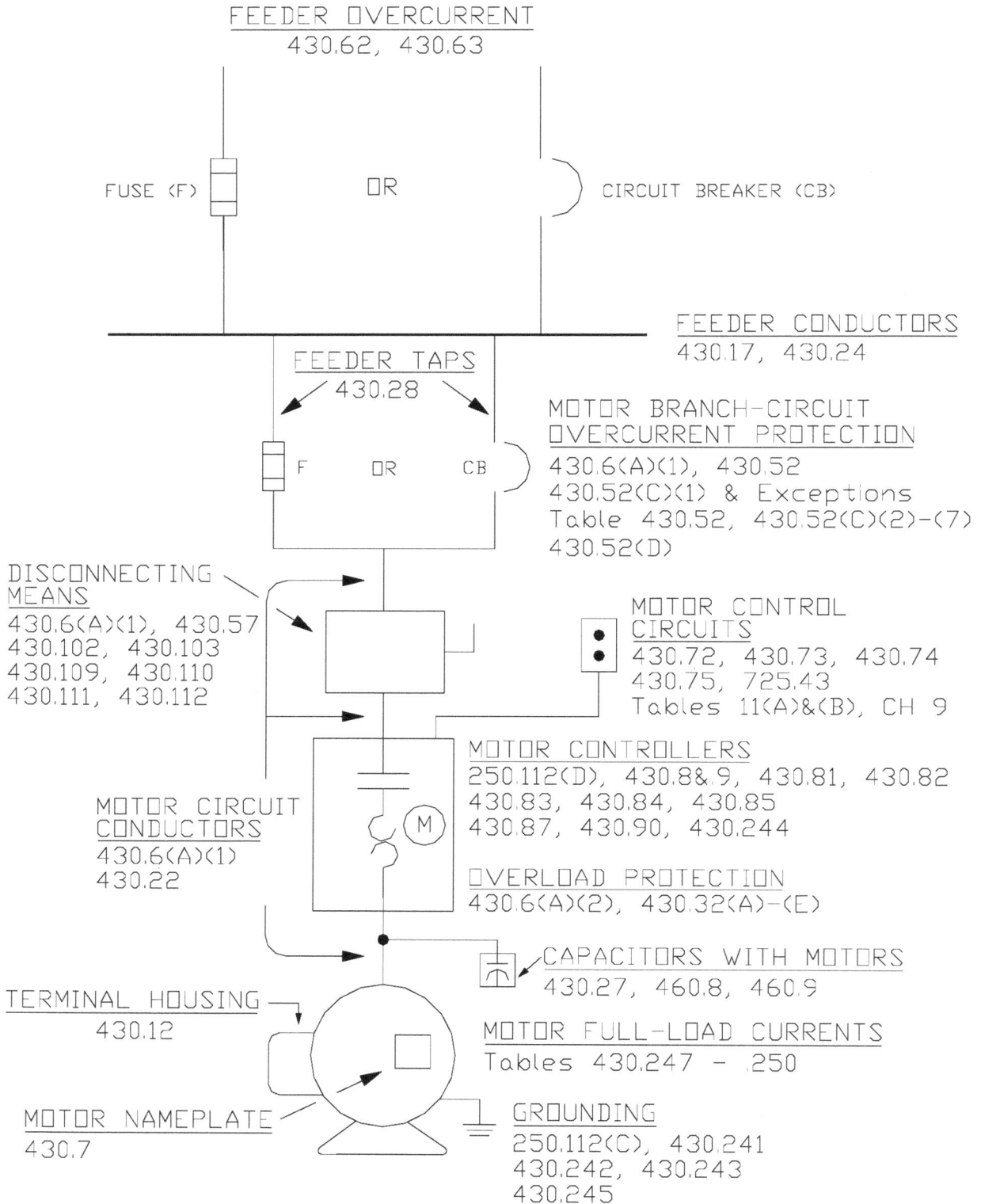

FEEDER OVERCURRENT
430.62, 430.63

FUSE (F)   OR   CIRCUIT BREAKER (CB)

FEEDER CONDUCTORS
430.17, 430.24

FEEDER TAPS
430.28

F   OR   CB

MOTOR BRANCH-CIRCUIT
OVERCURRENT PROTECTION
430.6(A)(1), 430.52
430.52(C)(1) & Exceptions
Table 430.52, 430.52(C)(2)-(7)
430.52(D)

DISCONNECTING
MEANS
430.6(A)(1), 430.57
430.102, 430.103
430.109, 430.110
430.111, 430.112

MOTOR CONTROL
CIRCUITS
430.72, 430.73, 430.74
430.75, 725.43
Tables 11(A)&(B), CH 9

MOTOR CONTROLLERS
250.112(D), 430.8&.9, 430.81, 430.82
430.83, 430.84, 430.85
430.87, 430.90, 430.244

MOTOR CIRCUIT
CONDUCTORS
430.6(A)(1)
430.22

OVERLOAD PROTECTION
430.6(A)(2), 430.32(A)-(E)

CAPACITORS WITH MOTORS
430.27, 460.8, 460.9

TERMINAL HOUSING
430.12

MOTOR FULL-LOAD CURRENTS
Tables 430.247 - .250

MOTOR NAMEPLATE
430.7

GROUNDING
250.112(C), 430.241
430.242, 430.243
430.245

ARTICLE 430 - Motors, Motor Circuits, and Controllers

430.4 - Part-Winding Motors (1. - 2.) [2]

1. A standard part-winding start induction motor requires the use of separate overload devices. The motor has a name plate full-load current rating of 35.7A, a 1.2 service factor and a 30°C temperature rise. What size overload devices are needed for this motor?

According to NEC 430.4, where separate overload devices are used each half of the motor winding shall be individually protected in accordance with NEC 430.32 and 430.37 with a trip current being one-half of that specified. NEC 430.6(A)(2) states that separate motor overload protection shall be based on the motor's nameplate current rating.

Referring to NEC 430.32(A)(1), the motor's nameplate full-load current rating must be increased by 125 percent (1.25 - per service factor and temperature rise), where 35.7A x 1.25 = 44.63A. To determine the trip current of the overload devices, the calculated value must be multiplied by one-half (½). Therefore, 44.63A x ½ or 44.63A / 2 = 22.32A.

As a result, the overload devices must be sized at (or have a trip current of) 22.32A.

In the event these overload devices are not sufficient to start the motor or carry the load, overload devices of a higher size are permitted be used per NEC 430.32(C) thus resulting to the motor's nameplate current being increased by140 percent (1.40) where, 35.7A x 1.40 = 49.98A. Again, to determine the trip current of the overload devices, the calculated value must be multiplied by one-half (½) where, 49.98A x ½ or 49.98A / 2 = 24.99A.

In concluding, the overload devices are permitted to be sized at (or have a trip current of) 24.99A to allow the motor to start or to prevent nuisance tripping of the motor.

2. Let's assume that the motor in question No. 1. is a 460V, 25HP, three-phase motor. What size inverse time breaker or dual element (time-delay) fuses are required to protect the motor?

According to NEC 430.4, each motor-winding connection shall have branch-circuit short-circuit and ground-fault (overcurrent) protection rated at *not more than one-half* of that *specified by* NEC 430.52. NEC 430.6(A)(1) states that branch-circuit short-circuit and ground-fault protection shall be based on the motor's full-load current values given in Tables 430.247 - 430.250.

Referring to Table 430.250, because the assumed motor is a 460V, 25 horsepower, three-phase motor, the motor's full-load current is 34A.

Referring now to Table 430.52, the motor's full-load current must be increased by 250 percent (2.50) to determine the size inverse time breaker required and 175 percent (1.75) to determine the required size of the dual element (time-delay) fuses. However, as far as the dual element (time-delay) fuses are concerned, the *Exception* to NEC 430.4 limits the rating of dual element fuses to 150 percent of the motor's full-load current if both windings are protected and such protection allows the motor to start. Therefore,

$$34A \times 2.50 = 85A \text{ (inverse time breaker)} \qquad 34A \times 1.50 = 51A \text{ (dual element fuses)}$$

Because NEC 430.52(C)(1), *Exception No. 1* is covered under the provisions of NEC 430.52, a 90A inverse time breaker is permitted. As for the fuses, they are limited to 50A, the use of 60A fuses would exceed the 150 percent limitation. To determine the branch-circuit short-circuit and ground-fault protection per protective device, the ratings of the permitted devices must be multiplied by one-half (½). Therefore,

$$90A \times \tfrac{1}{2} \text{ or } 90A / 2 = 45A \quad and \quad 50A \times \tfrac{1}{2} \text{ or } 50A / 2 = 25A$$

Based on the calculations, to protect the motor, either a 45A inverse time circuit breaker or a set of 25A dual element (time-delay) fuses is required.

430.6(A)(1) - Table Values (Ampacity and Motor Determination)

3. List each motor's full-load current value based on given horsepower, power supply and power supply voltage according to Tables 430.247 - .250.

		Power Supply				**Power Supply Voltage**				
				(SC/WR)	(S-T)	(90)	(120)	(180)	(240)	(500)
HP	(DC)	1φ-AC	2φ-AC	3φ-AC	3φ-AC	115	230	460	575	2300
¼		○				5.8	2.9	--	--	--
½	○					6.8	5.4	3.4	2.7	--
¾				○		6.4	3.2	1.6	1.3	--
1			○			6.4	3.2	1.6	1.3	--
2					○	--	--	--	--	--
3				○		--	9.6	4.8	3.9	--
5			○			--	28	--	--	--
7½	○					--	58	--	29	13.6
10		○				--	24	12	10	--
20				○		--	54	27	22	--
25					○	--	53	26	21	--
40			○			--	90	45	36	--
50		○				--	--	--	--	--
75					○	--	155	78	62	15
100				○		--	248	124	99	26

SC/WR – Squirrel Cage/Wound Rotor   S-T – Synchronous-Type

It is important to remember that the Full-Load Current Tables of Article 430 reflects the *worst-case* full-load current values of the various motors listed compared to actual operating current values.

430.6(A)(1), *Exception No. 2* - Table Values (Ampacity and Motor Rating Determination)

4. Exhaust fans are being installed in the restroom facilities of a commercial building. The fans are equipped with 1 horsepower, single-phase, 120-volt shaded-pole motors where the motor's terminal rating is 75°C. The nameplate full-load current of the motors is 12.8A. Determine the ampacity of the conductors required to supply each exhaust fan.

As it pertains to the rule, NEC 430.6(A)(1) states that the ampacity of a conductor supplying a motor must be determined based on the applicable full-load current values given in Tables 430.247 through 430.250.

However, just as there are rules, there are also *exceptions* to the rule as in this situation. NEC 430.6(A)(1), *Exception No. 2* states where shaded-pole or permanent-split capacitor-type fan or blower motors are used, the full-load current that is marked on the nameplate of the motor must be used instead of the horsepower rating to determine the ampacity of a branch-circuit conductor. In other words, the full-load current values given in Tables 430.247 through 430.250 are not used.

Using the given nameplate full-load current of the motors, 12.8A and NEC 430.22 which requires the ampacity of a branch circuit to be increased by 125 percent (1.25), the conductors supplying each exhaust fan motor must have an ampacity that is not less than 16A where,

$$12.8A \times 1.25 = 16A$$

Referring to Table 310.15(B)(16), at 75°C the minimum sized conductors required to supply these motor loads based on an ampacity of 16A must be 14 AWG copper conductors.

Just remember, if the sizing of overcurrent devices were required for these motors NEC 240.4(D) would not apply, that is, the ampere rating of the overcurrent device would not be determined based on the size of the conductor (14 AWG conductor - 15 amps device). NEC 240.4(D) reference NEC 240.4(G) which states that overcurrent protection for specific conductors shall be permitted to be provided as referenced in Table 240.4(G). Table 240.4(G) reference the use of Parts III - VII of Article 430 for sizing overcurrent protection for motors and motor-control circuit conductors.

430.6(B) - Torque Motors (Ampacity and Motor Rating Determination)

5. The nameplate of a 125HP, 3φ, 460 volt torque motor list the starting current of the motor at 928 amps. The letter G identifies the motor's locked rotor (LR) kVA Code. Determine the ampacity of the conductors that are required to supply this motor. Also see question Nos. 40. and 52. for sizing torque motor's overcurrent protection and disconnecting means.

For torque motors, the rated current shall be the locked-rotor current, and this nameplate current shall be used to determine the ampacity of the branch-circuit conductors according to NEC 430.6(B).

The motor's nameplate list the starting current which is in terms the same as the motor's locked rotor current as 928 amps. Therefore, based on 928 amps and the 125 percent requirement of NEC 430.22 the ampacity of the branch-circuit conductors is calculated where,

$$928A \times 1.25 = 1160A$$

As a minimum, the motor's branch-circuit conductors must have an ampacity that is at least 1160 amps. If the starting current had not been listed, the LR kVA Code letter could have been used instead to calculate the motor's locked rotor current per Table 430.7(B). See question No. 7.

430.6(C) - Alternating-Current Adjustable Voltage Motors (Ampacity and Motor Rating Determination)

6. A 15 HP, 208V 3-phase motor is used in accordance with NEC 430.6(C). The motor's nameplate only reflects the information as stated. Based on the information provided, determine the required ampacity of the branch-circuit conductors needed to supply the motor.

NEC 430.6(C) states that if the maximum operating current does not appear on the nameplate, the ampacity is determined based on the application of a 150 percent (1.50) increase of the applicable full-load current value found in either Table 430.249 or 430.250 that applies to the motor in use. In this case Table 430.250 will be referenced because the motor in use is a 3 phase motor. Looking at the table, a 15HP, 208V 3-phase motor has a full-load current of 46.2A. However, this is not the current value that can be used towards determining the needed ampacity, it's only a portion. The needed ampacity is obtained by increasing the motor's full-load current by 150 percent. Therefore,

$$46.2A \times 1.50 = 69.3A$$

With this calculated value we now have means to select the minimum size branch-circuit conductors needed to supply the motor along with means to make other determinants per NEC 430.6(C).

430.7(B) - Locked-Rotor Indicating Code Letter (Marking on Motors and Multimotor Equipmt.)

7. Determine the locked-rotor current for a three-phase 5 horsepower motor marked with the code letter **E** rated for 240V.

NEC 430.7(B) requires code letters that are marked on the nameplate of a motor to show the motor's input with locked rotor in accordance with Table 430.7(B). Table 430.7(B) provides a list of code letters which corresponds to the Kilovolt-Amperes (KVA) per horsepower with locked rotor. Notice how the values are listed in the table ranging from minimum to maximum values. For code letter E, the Kilovolt-Amperes range from 4.5 - 4.99. With this information and the formula provided below, the motor's lock-rotor current can be calculated.

$$LRC = \frac{kVA \text{ [listing per Table 430.7(B)] x 1000 x HP}}{V \times \sqrt{3}, \text{ where} \sqrt{3} = 1.732}$$

In solving this problem the maximum value (worst-case) will be used.

$$\frac{4.99kVA \times 1000 \times 5HP}{240V \times 1.732} = 60.0A$$ (As a reminder, 4.99kVA = 4990VA [k = 1000])

Applying the maximum locked-rotor value the locked-rotor current for the 3-phase, 5HP is 60A.

Table 430.10(B) - Minimum Wire-Bending Space at the Terminals of Enclosed Motor Controllers (8. - 9.) [2]

8. A 150 horsepower, three-phase, 460-volt motor is being fed from 3 - 4/0 AWG copper conductors via a combination starter. How much wire bending space is required to terminate the conductors in the combination starter?

Referring to Table 430.10(B), based on three 4/0 AWG single conductors terminating in three individual wire terminals, the minimum wire bending space for this installation must be 7".

9. If three 4/0 AWG parallel conductors (per phase) where used to supply the motor in question No. 8. instead, how much wire bending would be required to terminate the conductors?

The footnote beneath Table 430.10(B) states where provision for three or more wires per terminal exists, the minimum wire-bending space shall be in accordance with the requirements of Article 312.

Referring to Tables 312.6(A) and 312.6(B) where Table 312.6(A) per NEC 312.6(B)(1) is used to determine terminal space based on *conductors not entering or leaving (an) opposite wall* and Table 312.6(B) per NEC 312.6(B)(2) is used to determine terminal space based on *conductors entering or leaving (an) opposite wall.*

Because the question does not state how the conductors will enter and exit the combination starter we'll consider both conditions per **Tables 312.6(A)** and **B**.

Per Table 312.6(A), the first entry that allows 3-4/0 AWG wires (conductors) per terminal should be used to determine the wire bending space if the motor conductors will not enter or leave opposite walls of the enclosure of the combination starter. As a result, the minimum wire bending space for this installation must be 8".

Per Table 312.6(B), where 3 wires (conductors) per terminal are being installed, the minimum wire bending space must be no less than 8½" where the motor conductors will enter or leave the opposite walls of the enclosure of the combination starter.

Table 430.12(B) - Terminal Housings - Wire-to-Wire Connections [Motor 275 mm (11 in.) in Diameter or Less] (10. - 12.) [3]

10. Determine the minimum usable volume for the terminal housing of a 7.5 horsepower motor that measures 10.75" in diameter.

Per Table 430.12(B) the minimum usable volume for the terminal housing of such motor is required to be 22.4 in^3.

11. Determine the minimum usable volume for the terminal housing of a 12 lead, 200 HP, 3-phase, 460 volt motor that measures 28" in diameter.

Based on Table 430.250, a 200 HP, 3-phase, 460-volt motor has a full-load current of 240A which is less than the maximum full-load current (250A) per Table 430.12(B) where three-phase motors can employ up to 12 leads.

As a result, the terminal housing for this motor must have a 450 in^3 usable volume as a minimum.

12. Determine the minimum usable volume for the terminal housing of a 6 lead DC motor where the motor's full-load current is 283A.

The minimum usable volume for this motor's terminal housing must be determined based on the motor's full-load current (283A) and selected per maximum full-load current (375A) for DC motors with 6 leads as referenced in Table 430.12(B) which results to 330 in^3.

Table 430.12(C)(1) - Terminal Spacings - Fixed Terminals

13. The terminal housing of a 575V motor is equipped with permanently mounted line terminals. How much space is required between the line terminals? How much space is required between the line terminals and the terminal housing if uninsulated?

Per Table 430.12(C)(1), based on the voltage rating of the motor, the space required between the line terminals must be a minimum of 3/8" and the space required between the line terminals and an uninsulated terminal housing must be a minimum of 3/8" also.

Table 430.12(C)(2) - Usable Volumes - Fixed Terminals

14. Three 6 AWG conductors are used to energize a motor. The motor's terminal housing is equipped with fixed terminals. Determine the minimum usable volume of the terminal housing.

As referenced in Table 430.12(C)(2), the cubic inch capacity for a 6 AWG power-supply conductor is 2¼ (2.25) in^3. Therefore, the minimum usable volume of the terminal housing must be 6.75 (6¾) in^3 that is, 2.25in^3 x 3.

430.17 - Highest Rated or Smallest Rated Motor

15. The motors listed below are being fed from a 208/120V, 3φ source. Determine the highest and lowest rated motor based on the following motor data:

208 volts - 1φ	208 volts - 3φ	120 volts - 1φ
1 HP - 8.8A FLC	1 HP - 4.6A FLC	1 HP - 16A FLC
3 HP - 18.7A FLC	3 HP - 10.6A FLC	3 HP - 34A FLC
5 HP - 30.8A FLC	5 HP - 16.7A FLC	5 HP - 56A FLC
10 HP - 55.0A FLC	10 HP - 30.8A FLC	10 HP - 100A FLC

FLC = Full-Load Current

According to NEC 430.17, the highest or smallest rated motor shall be based on the rated-load current as selected from Tables 430.247, 430.248, 430.249 and 430.250 of Article 430.

Under most situations, if one was asked such a question without having knowledge of NEC 430.17, their answer would for certain be based on either the horsepower rating or the horsepower rating per supply voltage of a motor for determining the highest or smallest rating of a motor.

Based on the given motor data per Tables 430.248 (1φ) and 430.250 (3φ) the highest rated motor would be the 10 HP, 1φ, 120 volt motor which has a full-load current of 100A and the smallest rated motor would be the 1HP, 3φ, 208 volt motor which has a full-load current of 4.6A.

As you can see the horsepower rating, neither the number of phases nor the voltage rating of a motor are the determining factors as to how a motor is rated based on highest or smallest rating, only the motor's full-load current matters.

NEC 440.7 covers the highest rated (largest) motor as it pertains to air-conditioning and refrigeration equipment.

MOTOR CIRCUIT CONDUCTORS

ACCORDING TO THE NEC

The following NEC sections and tables are applicable for understanding how to size motor circuit conductors per Part II of Article 430.

NEC 430.22 - Single Motor - Conductors (branch-circuit) that supply single motors shall have an ampacity not less than 125 percent (1.25) of the motor's full-load current per NEC 430.6(A)(1), or not less than specified in NEC 430.22(A) - (G).

NEC 430.22(A) - Direct-Current Motor-Rectifier Supplied - Where *dc* motors operate from a rectified power supply, the conductor ampacity on the input of the rectifier shall not be less than 125 percent of the rated input current to the rectifier. Where *dc* motors operate from a rectified single-phase power supply, the conductors between the field wiring output terminals of the rectifier and the motor shall have an ampacity not less than 190 percent (1.90) for half-wave type rectifiers and not less than 150 percent (1.50) for full-wave type rectifiers. NEC 430.22(A) in the 2008 NEC is now NEC 430.22. NEC 430.22(A) is a new section in Article 430.

NEC 430.22(B) - Multispeed Motor - When multispeed motors are used, the ampacity of the branch-circuit conductors on the line side of the controller (from disconnecting means to controller) shall be based on the highest of the full-load current ratings shown on the motor nameplate and not according to NEC 430.6(A)(1). The ampacity of the branch-circuit conductors between the controller and the motor shall not be less than 125 percent of the current rating of the winding(s) that the conductors energize.

NEC 430.22(C) - Wye-Start, Delta-Run Motor - When wye-start, delta-run connected motors are used, the ampacity of the branch-circuit conductors on the line side of the controller (from disconnecting means to controller) shall not be less than 125 percent of the motor full-load current per NEC 430.6(A)(1). The ampacity of the conductors between the controller and the motor shall not be less than 72 percent of the motor full-load current rating per NEC 430.6(A)(1). In the 2008 NEC, the requirement was 58 percent (.58) of the motor's full-load current which then had to be increased by 125 percent resulting to 72 percent (.58 x 1.25).

NEC 430.22(D) - Part-Winding Motor - When part-winding connected motors are used, the ampacity of the branch-circuit conductors on the line side of the controller (from disconnecting means to controller) shall not be less than 125 percent of the motor full-load current per NEC 430.6(A)(1). The ampacity of the conductors between the controller and the motor shall not be less than 62.5 percent of the motor full-load current rating per NEC 430.6(A)(1). In the 2008 NEC, the requirement was 50 percent (.50) of the motor's full-load current which then had to be increased by 125 percent resulting to 62.5percent (.50 x 1.25).

NEC 430.22(E) - Other Than Continuous Duty - Unless granted by the authority having jurisdiction, motor conductors used for short-time, intermittent, periodic, or varying duty applications shall have an ampacity of not less than the percentage of the motor nameplate current rating shown in Table 430.22(E).

Table 430.22(E) - Duty-Cycle Service - Provides percentage values to calculate the ampacity of a motor conductor per motor nameplate current based upon the motor operating in a duty status other than continuous.

Table 430.22(G) - Conductors for Small Motors - Conductors for small motors shall not be smaller than 14 AWG unless otherwise permitted in NEC 430.22(G)(1) or (G)(2).

Wound-Rotor Secondary

NEC 430.23(A) - Continuous Duty - When a wound-rotor motor is used for continuous duty, the ampacity of the conductors that connect the wound-rotor to its controller shall be rated at 125 percent (1.25) of the full-load secondary current of the motor.

NEC 430.23(B) - Other than Continuous Duty - If wound-rotor motors are not used for continuous duty, the ampacity of the conductors from the secondary must not be less than those specified in Table 430.22(E).

NEC 430.23(C) - Resistor Separate from Controller - When the controller is mounted separately from the secondary resistor, the percentage values (of full-load secondary current) listed in Table 430.23(C) must be used to calculate the ampacity of the conductors between the controller and the resistor.

Table 430.23(C) - Secondary Conductor - Provides percentage values for determining the ampacity of secondary resistor conductors per resistor duty classification.

APPLYING THE NEC

Branch-circuit conductors supplying single motors used in a continuous duty application are required to have an ampacity not less than 125 percent of the motor's full-load current or as directed otherwise. Once the ampacity of the motor's branch-circuit conductor is determined, Tables 310.15(B)(16) - 310.15(B)(21) are referenced to select the proper size conductor per operating conditions.

The provisions of NEC 110.14(C) must also be considered when determining the size of a motor's conductor. This section states that the temperature rating associated with the ampacity of a conductor must not exceed the temperature rating of any termination, conductor, or device.

However, conductors having a higher temperature rating than the terminals of equipment are permitted to be used for ampacity adjustments, corrections, or both.

The most common temperature ratings listed in the ampacity tables are 60°C(140°F), 75°C(167°F) and 90°C(194°F). Generally speaking, NEC 110.14(C)(1) states where temperature ratings are not listed and marked on equipment terminals, the following provisions must be followed:

(a) For circuits rated 100 amperes or less or equipment terminals sized for 14 AWG through 1 AWG conductors, the ampacity of such conductors must be limited to 60°C. Conductors having a higher temperature rating (75°C and 90°C) are allowed to be used; however the ampacity of such conductor must be limited to 60°C also. When the temperature rating of an equipment terminal is listed and identified, the ampacity of a conductor is so determined.

(b) For circuits rated over 100 amperes or equipment terminals sized for conductors larger than 1 AWG, the ampacity of such conductors must be limited to 75°C. Conductors having a higher temperature rating (90°C) are allowed to be used; however the ampacity of such conductor must also be limited to 75°C. Again, when the temperature rating of an equipment terminal is listed and identified, the ampacity of a conductor is so determined.

When sizing the branch-circuit conductors for single motors, the motor's full-load or applicable operating current must be known. Once determined, the 125 percent increase can be applied based on NEC 430.22. Remember NEC 110.14(C)(1) must always be considered. Refer to the following procedures,

1. Determine the motor's full-load or operating current value per NEC.

2. Multiply the current value by 125 percent (1.25) to determine the ampacity of the motor's branch-circuit conductors.

3. Refer to the applicable ampacity table to size the motor's branch-circuit conductors.

4. If the terminal (temperature) rating of utilized equipment is unknown, limit the ampacity of the branch-circuit conductors per NEC 110.14(C)(1)(a) or (b), otherwise the ampacity of the branch-circuit conductors is determined based on the terminal rating of the utilized equipment.

430.22 - Single Motor (16. - 18.) [3]

16. What size THWN copper conductors are required to supply a 480V, three-phase, 50HP squirrel cage induction motor? The terminal rating of the equipment is unknown.

Question Nos. 16. and 17. will be used as examples to demonstrate the procedure for sizing the branch-circuit conductors for single motors.

Application

Step 1: Determine the motor's full-load or operating current value per NEC.

NEC 430.6(A) requires the values given in Tables 430.247 - 430.250 to be used in determining the ampacity of conductors supplying motors. Referring to Table 430.250, the full-load current for such motor is 65A.

Step 2: Multiply the current value by 125 percent (1.25) to determine the ampacity of the motor's branch-circuit conductors.

NEC 430.22 requires a motor's full-load current to be increased by 125 percent (1.25) to determine the ampacity of the motor's branch-circuit conductors. Unless otherwise, all motors are assumed to operate in a continuous duty application. Applying NEC 430.22,

$$65A \times 1.25 = 81.25A$$

Step 3: Refer to the applicable ampacity table to size the branch-circuit conductors.

Referring to Table 310.15(B)(16), based on an ampacity of 81.25A, 4 AWG THWN copper conductors rated for 75°C are required to be used as the motor's branch-circuit conductors.

Step 4: If the terminal (temperature) rating of utilized equipment is unknown, limit the ampacity of the branch-circuit conductors per NEC 110.14(C)(1)(a) or (b), otherwise the ampacity of the branch-circuit conductors is determined based on the terminal rating of the utilized equipment.

Because the terminal rating of the equipment (equipment meaning circuit breakers, fuses, terminal lugs, etc.) which will serve as the terminating means for the motor's branch-circuit conductors is unknown, the ampacity of the branch-circuit conductors must be determined based on 60°C instead of 75°C per NEC 110.14(C)(1)(a). Therefore, referring to Table 310.15(B)(16), 3 AWG THWN conductors, as a minimum, which have a rated ampacity of 100A must be used but limited to 85A at 60°C which will accommodate the 81.25A load.

17. The terminals of two three-phase 200A fusible disconnects are rated for 75°C. The disconnects will be used to provide protection for a three-phase, 230V, 50HP synchronous motor with a 90 percent power factor and a three-phase, 460V, 100HP synchronous motor with an 80 percent power factor. What size THWN-2 copper conductors are needed to supply the motors?

Application

Step 1: Determine each motor's full-load or operating current value per NEC.

Referring to Table 430.250, the full-load current for a three-phase, 230V, 50HP synchronous motor is 104A whereas the full-load current for a three-phase, 460V, 100HP synchronous motor is 101A. However, because the information provided in the footnote to Table 430.250 states that the full-load current listed in the table requires being increased 1.1 times when a synchronous motor has a power factor of 90 percent and 1.25 times when a synchronous motor has a power factor of 80 percent; these additional requirements must be considered. The full-load currents for such motors are,

$$104A \times 1.1 = 114.4A \text{ (60HP)}$$
$$\text{and}$$
$$101A \times 1.25 = 126.25A \text{ (100HP)}$$

Step 2: Multiply the current value by 125 percent (1.25) to determine the ampacity of the motor's branch-circuit conductors.

Both calculated current values per footnote to Table 430.250 are now increased by 125 percent (1.25) per NEC 430.22 to determine the required ampacity for each motor's branch-circuit conductors.

$$114.4A \times 1.25 = 143A \text{ (60HP)}$$
$$\text{and}$$
$$126.25A \times 1.25 = 157.81A \text{ (100HP)}$$

Step 3: Refer to the applicable ampacity table to size the branch-circuit conductors.

Based on the calculated ampacities of 143A and 157.81A respectively, a set of 1/0 AWG THWN-2 copper conductors with a rated ampacity of 150A at 75°C are required for the 50HP motor and a set of 2/0 AWG THWN-2 copper conductors with a rated ampacity of 175A at 75°C are required for the 100HP motor according to Table 310.15(B)(16) for use as the motor's branch-circuit conductors.

Step 4: If the terminal (temperature) rating of utilized equipment is unknown, limit the ampacity of the branch-circuit conductors per NEC 110.14(C)(1)(a) or (b), otherwise the ampacity of the branch-circuit conductors is determined based on the terminal rating of the utilized equipment.

NEC 110.14(C)(1)(b) states that the terminals of equipment, can only be used for conductors which are rated for 75°C when circuits (conductors) to be connected to such terminals are rated for over 100 amperes. This section also states that when the terminals of equipment are marked to allow the connection of conductors which are larger than 1 AWG, only conductors which are rated for 75°C can be used or where conductors with a higher temperature rating (90°C) is used the ampacity of the conductor must be limited to 75°C. In this situation, the THWN-2 copper conductors are rated for 90°C but are limited to an ampacity at 75°C.

18. Refer to Figure 430.22-18. What size THWN copper branch-circuit conductors (bcc) are required to supply each motor? The branch-circuit conductors will terminate in equipment rated for 75°C. The motors will be supplied from a 240/120V, 3-phase, 4W feeder. Question No. 18. is the initial question relating to Figure 430.22-18. Motor applications pertaining to Figure 430.22-18 are progressively resolved in six remaining questions, question Nos. 25., 32., 35., 38., 41. and 43.

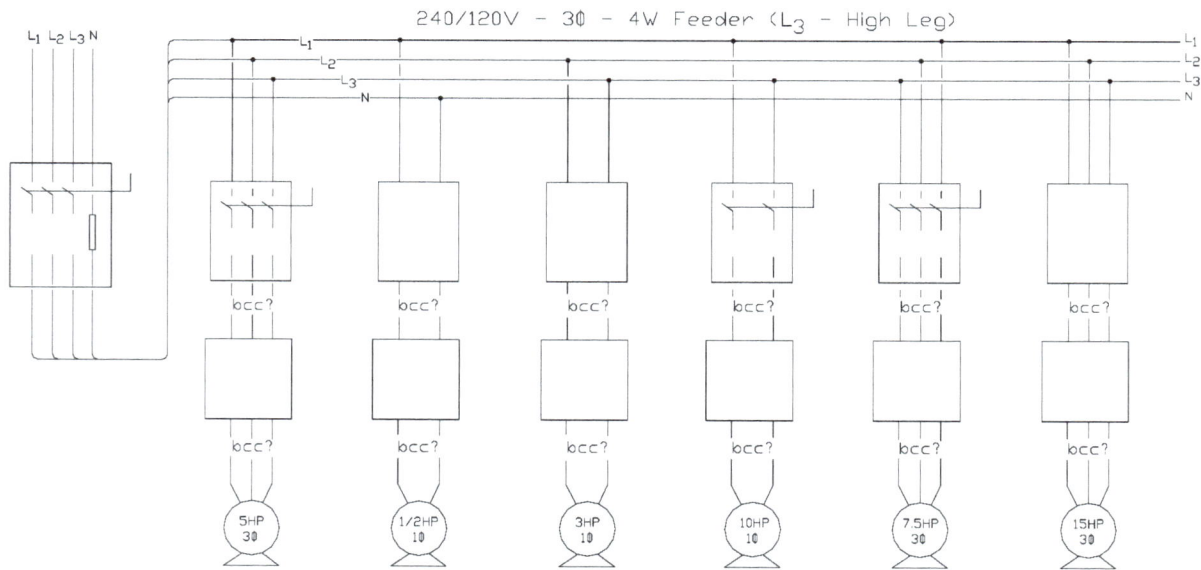

Figure 430.22-18

Referring to **Table 430.248**, the following full-load currents are provided for each single-phase motor.

Horsepower Rating	Full-load current
½	9.8A
3	17.0A
10	50.0A

Referring to **Table 430.250**, the following full-load currents are provided for each three-phase motor.

Horsepower Rating	Full-load current
5	15.2A
7.5	22.0A
15	42.0A

The ampacity for each set of branch-circuit conductors is as calculated,

5 HP motor - 15.2A x 1.25 = 19.0A   10 HP motor - 50.0A x 1.25 = 62.5A
½ HP motor - 9.8A x 1.25 = 12.25A   7.5 HP motor - 22.0A x 1.25 = 27.5A
3 HP motor -17.0A x 1.25 = 21.25A   15 HP motor - 42.0A x 1.25 = 52.5A

The following THWN copper conductors @ 75°C are required to supply the motors according to the ampacities listed in Table 310.15(B)(16).

5 HP motor at 19.0A, use a set of 14 AWG conductors rated for 20A
½ HP motor at 12.25A, use a set of 14 AWG conductors rated for 20A
3 HP motor at 21.25A, use a set of 12 AWG conductors rated for 25A
10 HP motor at 62.5A, use a set of 6 AWG conductors rated for 65A
7.5 HP motor at 27.5A, use a set of 10 AWG conductors rated for 35A
15 HP motor at 52.5A, use a set of 6 AWG conductors rated for 65A

430.22(A) - Direct-Current Motor-Rectifier Supplied

Simply said, a *rectifier* is a device that converts alternating current (*ac*) that alternately flows in two directions to a direct current (*dc*) that only flows in one direction. The device may consist of a single diode (an electrical component that only allows current to flow in one direction) or a combination of diodes, depending upon the need of rectification. Either way, rectification can be accomplished by either a *half* or *full-wave* rectifier. Where a *half-wave* rectifier is used, only the positive or negative half of an *ac* cycle is allowed to pass while the other half of the cycle is blocked. If a half-wave rectifier is used in a circuit, the circuit is referred to as a *half-wave* circuit. On the hand, where a *full-wave* rectifier is used, the full *ac* cycle is allowed to pass which includes both halves of the cycle. If a full-wave rectifier is used in a circuit, the circuit is referred to as a *full-wave* circuit.

19. A group of *dc* motors are supplied from a three-phase rectified power supply. Determine the required ampacity of the supplying conductors if the input current to the power supply is 137.4A. Also determine the conductor ampacity if a single-phase half or full-wave rectifier is used.

According to NEC 430.22(A), where dc motors operate from a rectified power supply, the conductor ampacity on the input of the rectifier shall not be less than 125 percent of the rated input current to the rectifier. As a result, the ampacity of the supplying conductors are required to have an ampacity not less than 171.75A (137.4A x 1.25).

As for a single-phase half-wave rectifier being used, NEC 430.22(A)(1) requires the ampacity of the conductors to be not less than 190 percent resulting to an ampacity not less than 261.06A (137.4A x 1.90) whereas, for a single-phase full-wave rectifier being used, NEC 430.22(A)(2) requires the ampacity of the conductors to be not less than 150 percent resulting to an ampacity not less than 206.1A (137.4A x 1.50).

430.22(B) - Multispeed Motor

20. Refer to Figure 430.22(B)-20. What size copper branch-circuit conductors are required for the multispeed motor that runs from the disconnecting means to the line side of the controller and from the controller to the motor? The terminal rating of the disconnecting means and controller is rated for 60°C.

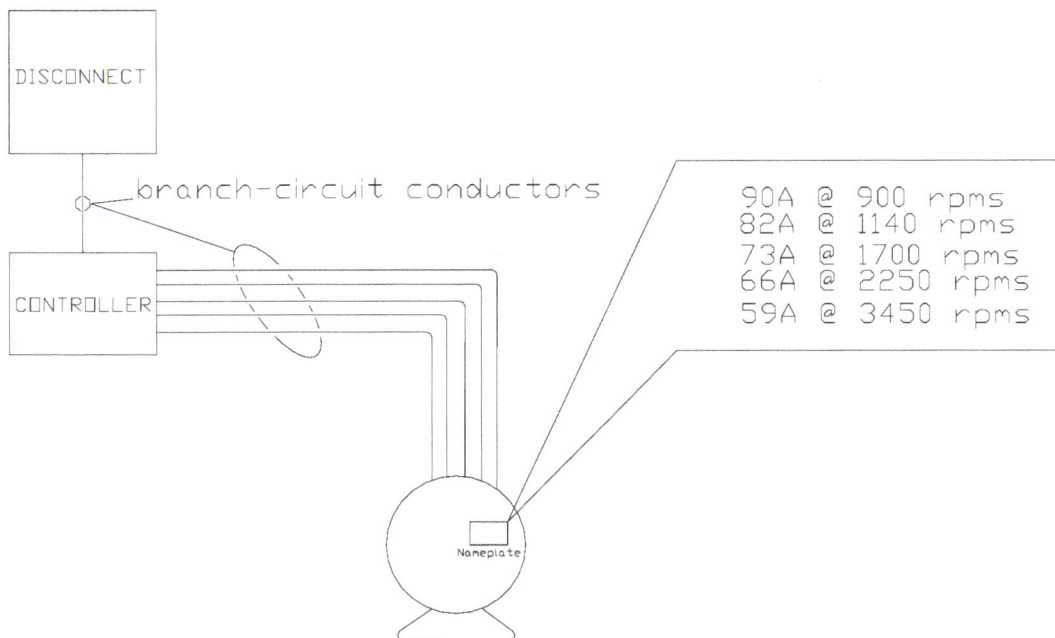

DISCONNECT

branch-circuit conductors

90A @ 900 rpms
82A @ 1140 rpms
73A @ 1700 rpms
66A @ 2250 rpms
59A @ 3450 rpms

CONTROLLER

Nameplate

Figure 430.22(B)-20

According to NEC 430.22(B) the motor's operating current is not determined by the table values but is based on the nameplate current shown on the nameplate of the motor. As shown in Figure 430.22(B)-20, the motor's nameplate is marked with the full-load current values per speed variation. Therefore, to size the branch-circuit conductors on the line side of the controller, NEC 430.22(B) states that the branch-circuit conductors must be sized based on the highest full-load current value marked on the nameplate of the motor which is 90A at 900 rpms. This current value reflects the amount of current the motor's highest winding will experience under full load at the given speed.

As for the conductors extending from the controller to the motor, individual branch-circuit conductors must be provided per motor winding which corresponds to the desired speed of the motor. Because this is a five speed motor the motor's nameplate is also marked with the remaining full-load current values; 82A at 1140 rpms, 73A at 1700 rpms, 66A at 2250 rpms and 59A at 3450 rpms.

NEC 430.22 must be applied for each individual full-load current value. The ampacity for the branch-circuit conductors that runs from the disconnecting means to the line side of the controller is as calculated,

$$90A \times 1.25 = 112.50A$$

The ampacity per motor winding's current rating to size the branch-circuit conductors (per speed) that runs from the controller to the motor is as calculated,

$$82A \times 1.25 = 102.50A \quad 73A \times 1.25 = 91.25A$$
$$66A \times 1.25 = 82.50A \quad 59A \times 1.25 = 73.75A$$

The calculated ampacity at 112.50A must again be applied with the above ampacities because such related conductors will also be run from the controller to the motor.

Based on the ampacities listed in the 60°C column of Table 310.15(B)(16) for copper conductors, the following conductors are required per calculated ampacities,

branch-circuit conductors - disconnecting means to line side of controller
at 112.50A use 1/0 AWG copper (125A)

branch-circuit conductors - load side of controller to motor
at 112.50A use 1/0 AWG copper (125A)
at 102.50A use 1 AWG copper (110A)
at 91.25A use 2 AWG copper (95A)
at 82.50A use 3 AWG copper (85A)
at 73.75A use 3 AWG copper (85A)

Use a set of 1/0 AWG copper conductors for the branch-circuit conductors supplying the line side of the controller and from the controller to the motor. Use copper conductors of sizes 1 AWG, 2 AWG, and 3 AWG (twice) respectively, per motor speed for branch-circuit conductors being run from the controller to the motor.

430.22(C) - Wye-Start, Delta-Run Motor *and* **430.22(D)** - Part-Winding Motor

21. Refer Figure 430.22(C)/(D)-21. What size copper branch-circuit conductors are required to supply each individual motor? The terminal rating of each motor's overcurrent device and controller is rated for 75°C.

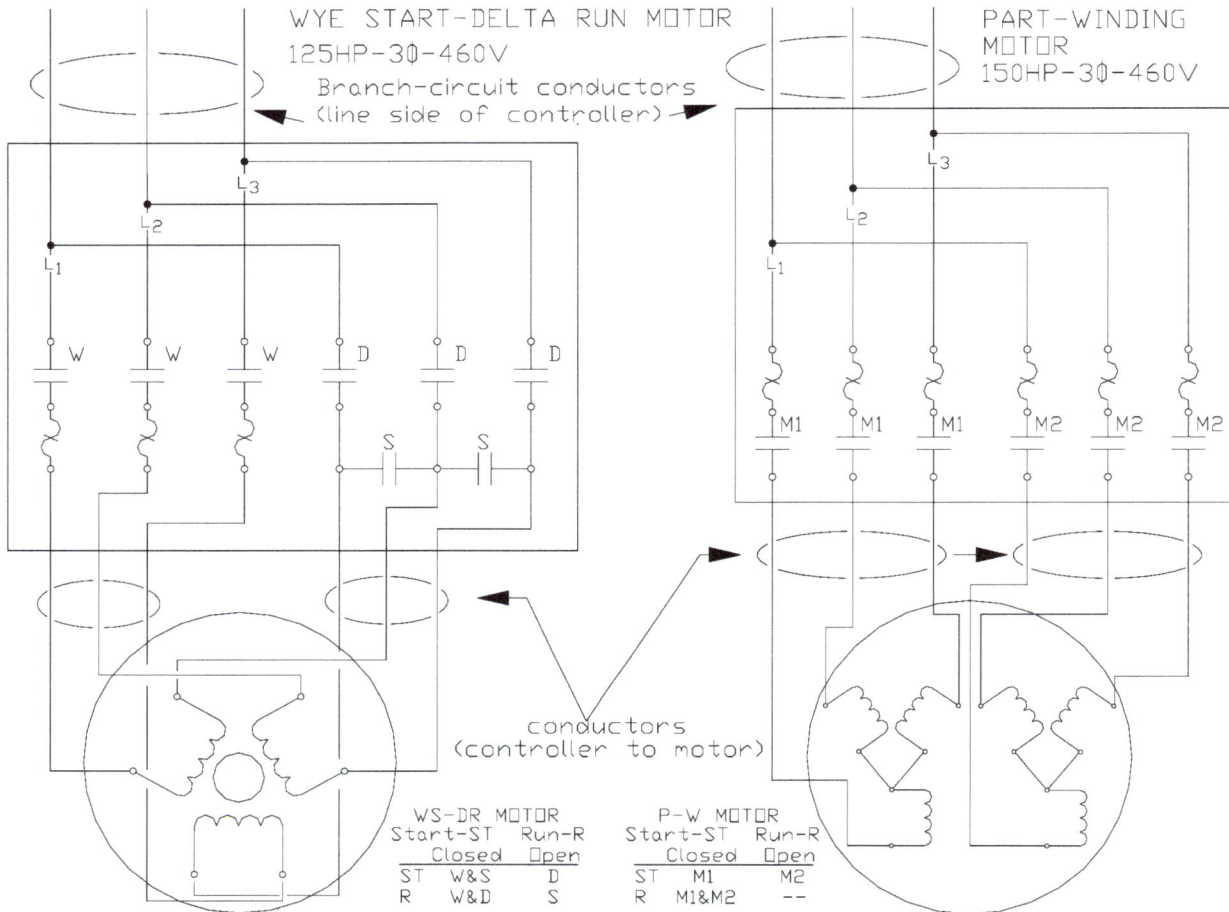

WYE START-DELTA RUN MOTOR
125HP-3∅-460V
Branch-circuit conductors
(line side of controller)

PART-WINDING
MOTOR
150HP-3∅-460V

conductors
(controller to motor)

WS-DR MOTOR		
Start-ST	Run-R	
Closed	Open	
ST	W&S	D
R	W&D	S

P-W MOTOR		
Start-ST	Run-R	
Closed	Open	
ST	M1	M2
R	M1&M2	--

Figure 430.22(C)/(D) - 21

According to NEC 430.22(C) and 430.22(D) the selection of branch-circuit conductors for wye-start, delta-run and part-winding connected motors on the line side of the controller is based upon the motor's full-load current. Referring to Table 430.250, the full-load currents for the 3-phase, 125HP and 150HP motors at 460V is 156A and 180A, respectively. As stated, the above full-load currents for both motors are used only for sizing the branch-circuit conductors on the line side of the controller.

The conclusion of both NEC 430.22(C) and 430.22(D) states that the selection of conductors between the controller and motor for a wye-start, delta-run connected motor is based on 72 percent (.72) of the motor's full-load current and 62.5 percent (.625) of the motor's full-load current for a part-winding connected motor.

As a result, the required ampacity of both type conductors is as follows:

Ampacity of branch-circuit conductors - line side of controller
wye-start, delta run - 180A x 1.25 = 225A
part-winding   - 156A x 1.25 = 195A

Ampacity of conductors - from controller to motor
wye-start, delta run* - 180A x .72   = 129.6A
part-winding   - 156A x .625 = 97.5A

*Motors that can run on either wye or delta draw one-third the starting current when wye-connected. When delta-connected the motor draws 3 times the starting current as when wye-connected.

Based on the ampacities listed in the 75°C column of Table 310.15(B)(16) for copper conductors, the following conductors are required per calculated ampacities,

branch-circuit conductors - line side of controller
at 225A use 4/0 AWG copper (230A)
at 195A use 3/0 AWG copper (200A)

conductors - line side of controller
at 129.6A use1/0 AWG copper (150A)
at 97.5A use 3 AWG copper (100A)

Because the terminal rating of the disconnecting means and controller were specified, the ampacity of the conductors were determined accordingly.

430.22(E) - Other Than Continuous Duty *and* **Table 430.22(E) - Duty-Cycle Service**

22. What size copper branch-circuit conductors are required to supply the *ac* induction motors below? The temperature rating of the terminating means is unknown.

Motor	Duty Cycle (Rated Time)*	HP	Volts	Phases	Nameplate Current
1	Short-time (15 min.)	15	575	3	14
2	Intermittent (5 min.)	40	460	3	45
3	Periodic (continuous)	60	230	3	146
4	Varying (30 and 60 min.)	10	208	1	44

*Time rating is required to be marked on motors per NEC 430.7(A)(6) in ratings of 5, 15, 30 or 60 minutes, or continuous. These ratings reference the amount of time a motor will be operated for example, a 5-minute rated motor will operate for 5 minutes and then will not operate with load for 55 minutes - likewise for a 15-minute rated motor (15 on/45 off) and for a 30-minute rated motor (30 on/30 off).

According to NEC 430.22(E) the operating current for motors having a duty-cycle other than continuous is not determined by Tables 430.248 (single-phase *ac* motors) and 430.250 (three-phase *ac* motors) but is based on the nameplate current shown on the nameplate of the motor and finally derived by the percentage values provided in Table 430.22(E). Referring to NEC 430.22(E) again, we find that the 125 percent increase as required by NEC 430.22 is replaced with the percentage values of Table 430.22(E) when determining the ampacity of conductors for motors with varying duty cycles.

Therefore, the nameplate current of each motor must be adjusted based on the percentage values listed in Table 430.22(E) to determine the ampacity of each motor's branch-circuit conductors.

Adjusted Nameplate Currents

Motor 1 [Short-time (15 min.)] - 14A x 1.20   =   **16.8A**

Motor 2 [Intermittent (5 min.)] - 45A x .85   =   **38.25A**

Motor 3 [Periodic (continuous)] - 146A x 1.40   = **204.4A**

Motor 4 [Varying (30 and 60 min.)] - 44A x 1.50   =   **66A**

The information provided in the original question only asked for the size copper conductors needed to supply the four motors. No information was provided pertaining to the copper conductors, that is, the insulation type or temperature rating of the conductors. However, the question concludes by stating, "the temperature rating of the terminating means is unknown".

Because the temperature ratings of the terminating means for the motor's branch-circuit conductors are unknown, the ampacity of the motor's branch-circuit conductors must be determined based on 60°C for circuits rated less than 100A per NEC 110.14(C)(1)(a). In this case, this would pertain to all motors with the exception of Motor 3. With this in mind, conductors rated for 60°C at their rated ampacity could be used or conductors rated for 75°C could also be used but at an ampacity limited to 60°C. For this question the latter will be used. If the question had identified the motors as being marked with design letters B, C, or D, the insulation rating of the branch-circuit conductors could have been higher (up to 90°C) providing the ampacity of the conductors do not exceed 75°C per NEC 110.14(C)(1)(a)(4). See NEC 110.14(C)(1)(a).

For Motor 3, the provisions of NEC 110.14(C)(1)(b) requires circuits rated over 100A to be limited to 75°C when the terminating means (for this motor's branch-circuit conductors) are unknown.

Based on the adjusted nameplate current of each motor, the motor's branch-circuit conductors will be sized according to the ampacities listed in Table 310.15(B)(16). Now that there are guidelines for sizing the motor's branch-circuit conductors let's do so.

<u>75°C conductors using 60°C ampacities</u>

Motor 1 at 16.8A will use 12 AWG copper conductors rated for 20A

Motor 2 at 38.25A will use 8 AWG copper conductors rated for 40A

Motor 4 at 66A will use 4 AWG copper conductors rated for 70A

<u>75°C conductors using 75°C ampacity</u>

Motor 3 at 204.4A will use 4/0 AWG copper conductors rated for 230A

430.23(A) - Continuous Duty *and* **430.23(C)** - Resistor Separate from Controller (Wound-Rotor Secondary)

23. Refer Figure 430.23(A)/(C)-23. The 3-phase, 75HP wound-rotor motor is rated for 480V. The motor's secondary rotor current per nameplate is 82A. The resistors in the secondary load bank are classified for medium intermittent duty. Using THWN copper conductors, size the motor's branch-circuit conductors, secondary conductors and resistor bank conductors.

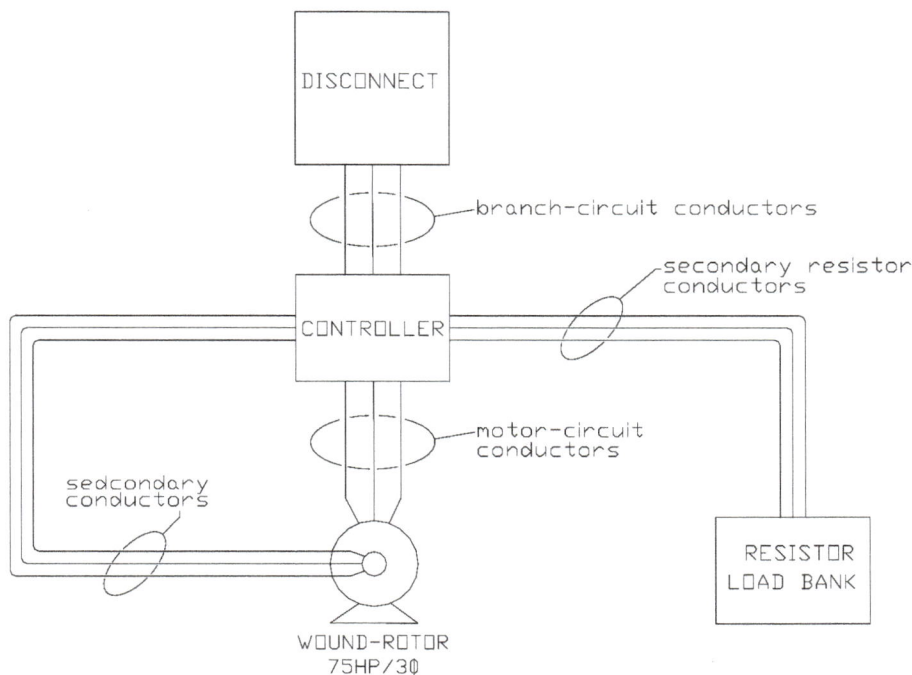

430.23(A)/(C)-23

Referring to Table 430.250, the full-load current for the wound-rotor motor is 96A and the motor's secondary rotor current as listed on the nameplate is 82A.

The ampacity for the branch and motor-circuit conductors is calculated per NEC 430.22,

$$96A \times 1.25 = 120A$$

In addition to determining the ampacity for the branch-circuit conductors, the ampacity must be determined to size the wound-rotor secondary circuit conductors and the secondary resistor conductors. In this case the nameplate current must be used to determine these ampacities instead of the full-load current per Table 430.250. The 125 percent adjustment of the "full-load secondary current of the motor" as stated in NEC 430.23(A), is referencing the motor's nameplate current. If not found on the motor's nameplate, this current rating could be provided by the manufacturer.

As a result, the ampacity for the secondary circuit conductors is as calculated, based on the given nameplate current,

$$82A \times 1.25 = 102.5A$$

Now to determine the ampacity for the secondary resistor conductors, NEC 430.23(C) requires the use of Table 430.23(C). Because the resistors in the secondary load bank are classified for medium intermittent duty, the ampacity of the conductors must be determined based on the percentage value provided in the table per given duty classification. Thus, the ampacity for the secondary resistor conductors is as calculated,

$$82A \times .75 = 61.5A$$

Referring to the 75°C column in Table 310.15(B)(16) which reference THWN copper conductors, a set of 1 AWG conductors can be used for the branch and motor-circuit conductors, a set of 2 AWG conductors can be used for the secondary circuit conductors, and a set of 6 AWG conductors can be used for the secondary resistor conductors.

Again notice how all conductors were sized based on having a temperature rating of 75°C. In this particular situation the conductors were determined based on insulation type, THWN. Needless to say, if the terminating means for these conductors are unknown you should know by now the required procedure for correcting such installations.

For an additional question pertaining to this subject matter, refer to question No. 3. of Article 610.

FEEDER CONDUCTORS

ACCORDING TO THE NEC

The following NEC sections are applicable for understanding how to size motor feeder conductors per Part II of Article 430.

NEC 430.24 - Several Motors or a Motor(s) and Other Load(s) - The ampacity of a feeder conductor that supply two or more motors shall have a sum not less than 125 percent (1.25) of the highest rated motor, per NEC 430.6(A) *plus* the sum of the full-load current ratings of all the motors in the group, per NEC 430.6(A) *plus* 100 percent of the noncontinuous non-motor load *plus* 125 percent of the continuous non-motor load.

> *Exception No. 1* - When one or more motors of a group of motors are rated for a duty cycle other than continuous duty (*i.e.* short-time, intermittent, periodic, or varying duty) the ampere rating of each motor is determined in accordance with NEC 430.22(E). The highest rated motor is determined based on the greater of either the ampere rating derived from NEC 430.22(E) or the largest continuous duty motor's full-load current multiplied by 1.25 (125 percent). Once the ampere rating or the increased full-load current of the highest rated motor is determined, all remaining motor's ampere ratings are added together.

> *Exception No. 2* - For a motor that is associated with a fixed electric space heating unit, the ampacity of the motor's conductors is based on NEC 424.3(B).

NEC 430.25 - Multimotor and Combination-Load Equipment - NEC 430.7 reference the ampacity of conductors that supplly multimotors and combination-load equipment where the ampacity of these conductors must not be less than the minimum ampacity marked on the equipment per NEC 430.7(D). When equipment is not factory-wired and individual nameplates are visible per NEC 430.7(D)(2), the conductor ampacity must be determined in accordance with NEC 430.24.

NEC 430.26 - Feeder Demand Factor - The authority having jurisdiction may grant special permission for feeder conductors to be of less ampacity than specified in NEC 430.24 where motors operate on duty-cycle, intermittently or where all motors do not operate at the same time. The conductors shall have sufficient ampacity for the maximum load determined in accordance with the sizes and numbers of motors supplied and the character of their loads and duties.

NEC 430.28 - Feeder Taps - Feeder tap conductors that are tapped from a feeder conductor are required to have an ampacity in accordance with Part II of Article 430, terminate in a branch-circuit overcurrent protection device and meet one of the following conditions:

(1) For feeder *tap* conductors not extending over 10 feet from the feeder conductors, the tap conductors must be enclosed by either a controller or raceway, *and* (for field installation) be protected by an overcurrent device on the line side of the tap conductors where the rating or setting of which shall not exceed 1000 percent (10 times) of the tap conductors ampacity *or*

(2) For feeder *tap* conductors extending over 10 feet and up to 25 feet from the feeder conductors, the ampacity of the tap conductors must not be less than one-third (1/3) of the feeder conductor's ampacity, *and* the tap conductor must be protected from physical damage or be enclosed in raceway *or*

(3) For feeder *tap* conductors not restricted to length, the ampacity of the tap conductors must be the same as the feeder conductors.

APPLYING THE NEC

FOR MOTOR LOADS ONLY

Feeder conductors supplying several motors are required to have an ampacity not less than 125 percent of the full-load current rating of the highest rated motor (as identified in NEC 430.17) *plus* the full-load current ratings of all other involved motors and nonmotor loads.

1. Determine each motor's full-load current value per NEC 430.6(A)(1).

2. Multiply the full-load current of the highest rated motor by 125 percent (1.25) and add the results to the full-load currents of all remaining motors per NEC 430.6(A)(1) plus non-motor noncontinuous loads (at 100 percent) and continuous loads (at 125 percent).

3. Refer to the applicable ampacity table to size the feeder conductors.

4. If the terminal (temperature) rating of affiliated equipment is unknown limit the ampacity of the feeder conductors per NEC 110.14(C)(1), otherwise the ampacity of the feeder conductors is determined based on the terminal rating of the equipment being utilized.

430.24 - Several Motors or a Motor(s) and Other Load(s) (24. - 28.) [5]

24. What size three-phase, 460V feeder conductors (THW copper) are required to supply the following group of three-phase motors: 1½HP, 3HP, 5HP, 7½HP(2), 10HP, 15HP, 25HP? The terminal rating of the equipment is 75°C.

Referring to Table 430.250, the listed full-load current value for each motor is:

Horsepower Rating	Full-load current
1½	3.0A
3	4.8A
5	7.6A
7½ (2)	11.0A
10	14.0A
15	21.0A
25	34.0A

Multiply the full-load current of the highest rated motor (per NEC 430.17) by 125 percent (per NEC 430.24) and add the results to the full-load currents of all remaining motors.

Highest Rated Motor = 25HP at 34A, therefore

$$34A \times 1.25 = 42.5A + (3A + 4.8A + 7.6A + [11A \times 2] + 14A + 21A) = 114.9A$$

Based on an ampacity of 114.9A, 2 AWG THW copper conductors rated for 75°C are required according to Table 310.15(B)(16).

25. Refer to Figure 430.24-25. Determine the minimum size of the feeder conductors where THWN copper conductors are required. The feeder conductors are connected to a fusible disconnect switch which has a terminal rating of 75°C. This question is a continuation of question No. 18.

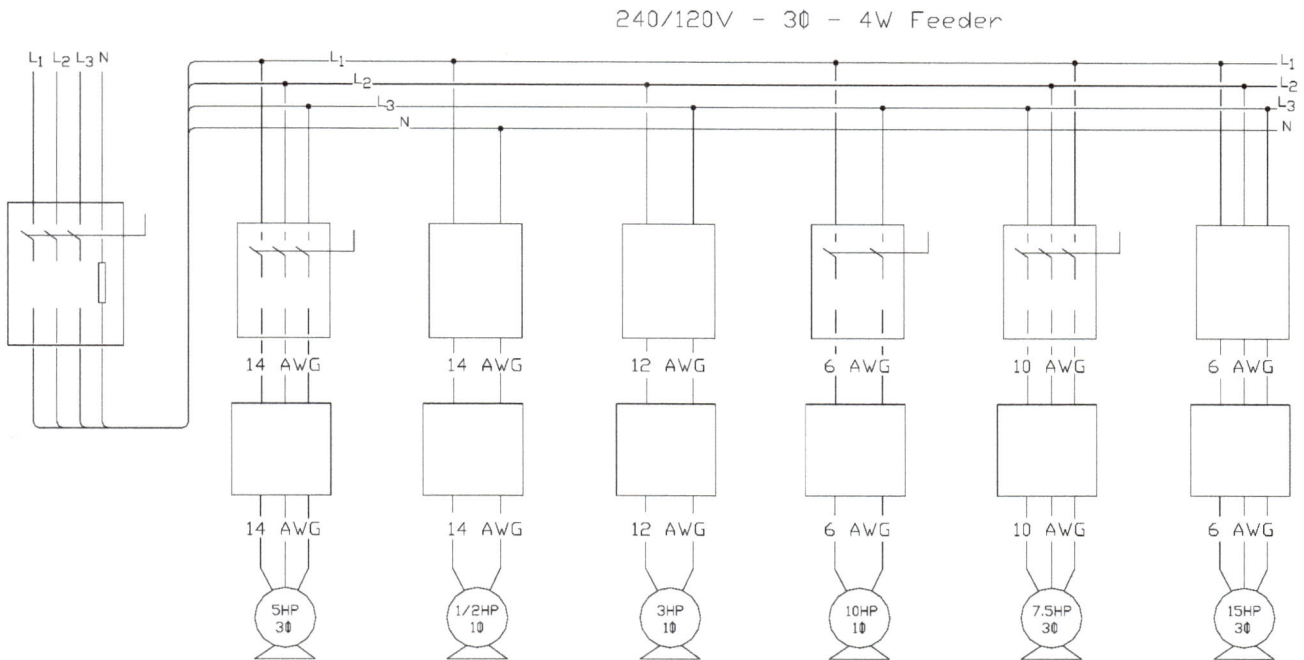

Figure 430.24-25

Determine each motor's full-load current value per NEC 430.248 and 430.250. Multiply the full-load current of the highest rated motor by 125 percent (1.25) and add the results to the full-load currents of all remaining motors.

By now it should be understood that the highest rated motor according to NEC 430.17 is based on the motor having the highest full-load current opposed to having the highest horsepower rating, the highest voltage or the highest number of phases. Although the 10HP single-phase motor is smaller in horsepower rating compared to the 15HP three-phase motor, this motor is rated the highest based on its full-load current. Therefore, deriving the required ampacity for the feeder conductors based on "all motors" amount to,

$$50A \times 1.25 = 62.5A + (9.8A + 15.2A + 17A + 22A + 42A) = 168.5A$$

The calculated ampacity alone can be used to size the feeder conductors for L_1, L_2 and L_3. As for as the neutral (grounded) conductor, NEC 215.2(A)(2) and 250.122 requires it to be sized no smaller than the provisions of Table 250.122. Nevertheless, the neutral (grounded) conductor must be sized to carry or exceed the neutral load and adhere to NEC 250.122. See the concluding paragraph of question No. 43.

As an alternative, the ampacity for each individual feeder conductor can be calculated "per line" to size each conductor without violating NEC 430.24.

Observe,

Feeder Conductor L_1

The largest rated motor connected to L_1 is the 10HP single-phase motor resulting to the following calculated ampacity for this conductor where,

$$50A \times 1.25 = 62.5A + (15.2A + 9.8A + 22A + 42A) = 151.5A$$

Feeder Conductor L_2

The largest rated motor connected to L_2 is now the 15HP three-phase motor which results to the following calculated ampacity for this conductor where,

$$42A \times 1.25 = 52.5A + (15.2A + 17A + 22A) = 106.7A$$

Feeder Conductor L3

The largest rated motor connected to L_3 is again the 10HP single-phase motor which results to the following calculated ampacity for this conductor where,

$$50A \times 1.25 = 62.5A + (15.2A + 17A + 22A + 42A) = 158.7A$$

Again, referring to Table 310.15(B)(16), the conductors required for feeders L_1 and L_3 are 2/0 AWG (175A) and for feeder L_2 a 2 AWG (115A). With the exception of feeder L_2, the same size conductors applying the alternative is approximately the same as if all feeder conductors were identically sized based on the calculated 168.5A determinant.

As a result, either three 2/0 AWG THWN copper conductors based on the consideration of "all motors" or the "per line" alternative combination of two 2/0 AWG and one 2 AWG THWN copper conductors can be used to serves as the feeder conductors supplying the motors. Either choice meets the minimum requirements. However, using the "per line" alternative limits the future expansion of the overall feeder source per line L_2.

See question No. 27. where the line conductors are sized based on the largest calculated line load.

26. What size feeder conductors are needed to supply a 25HP, 480V, 3-three phase motor along with two 37.5kVA, three-phase, 480Δ-208Y/120V transformers? Both transformers will be used to supply single phase loads. The equipment's terminating means for this installation is rated for 75°C.

The full-load current for the 25HP motor is 34A according to Table 430.250. Once the full-load current of the highest rated motor is increased by 125 percent, (34A x 1.25 = 42.5A) all remaining non-motor loads are then added to the motor load.

Transformers (Calculated at 100 percent of rating)

$$37.5kVA \, / \, 480V \times 1.732 = 45.1A \quad 45.1A \times 2 \text{ (transformers)} = 90.2A$$

TOTAL - motors + other loads - 42.5A + 90.2A = 132.7A

If the transformers were being used to supply continuous loads the required ampacity of the feeder conductors would have resulted to 155.25A (42.5A + [90.2A x 1.25]).

Because the type conductors were not specified we'll size the feeder conductors for both copper and aluminum according to Table 310.15(B)(16). The minimum size conductors required based on the calculated ampacity is,

copper (75°C) = 1/0 AWG (rated for 150A)   aluminum (75°C) = 2/0 AWG (rated for 135A)

27. What size THHN aluminum feeder conductors are needed to supply the loads shown in Figure 430.24-27? The temperature rating of the terminating means is unknown. All motors are rated for 240V. See Article 215 (Volume 2), question No. 4 (feeder supplying non-motor loads).

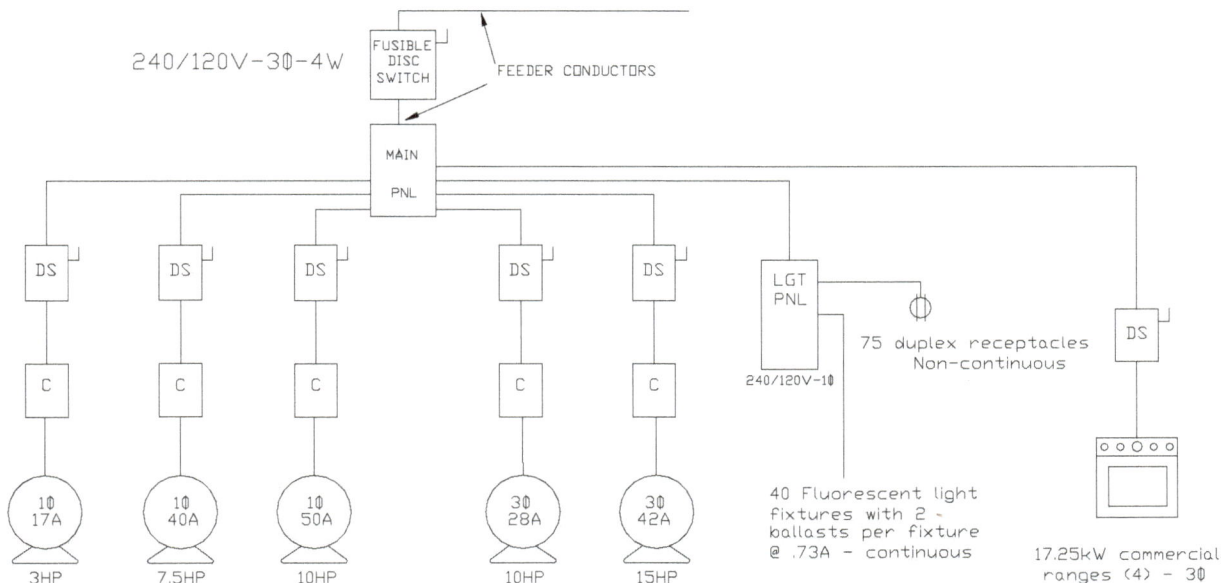

Figure 430.24-27

Referring to Tables 430.248 and 430.250, the following full-load currents are provided:

Horsepower Rating	Full-load current
3 (1φ)	17.0A
7.5 (1φ)	40.0A
10 (1φ)	50.0A (Highest Rated)
10 (3φ)	28.0A
15 (3φ)	42.0A

Multiply the full-load current of the highest rated motor by 125 percent (1.25) and add the results to the full-load currents of all remaining motors.

$$50A \times 1.25 = 62.5A + 17A + 40A + 28A + 42A = 189.5A$$

Add the ampacity of all other loads (non-motor) to the above results.

Considering all other loads,

Lighting (continuous) - 40 fixtures w/2 ballast per fixture rated for .73A each

$$40 \times 2 \times .73A \ (58.4A) \times 1.25 = 73A$$

$$73A/2 = \underline{36.5A} \text{ (per L-N)}$$

Receptacles (noncontinuous) - 75 duplex receptacles

$$75 \times 180VA = 13,500VA$$
[NEC 220.14(I), 220.44, Table 220.44]

$$
\begin{aligned}
\text{lst 10kVA} &= 10,000VA \\
\text{(Remaining) } 3500VA \times .50 &= \underline{1,750VA} \\
& \quad 11,750VA
\end{aligned}
$$

$$11,750VA \ / \ 240V = 48.96A$$

$$48.96A \ /2 = \underline{24.48A} \text{ (per L-N) [worst case]}$$

Kitchen Equipment (4) -17.25kW, 3φ commercial ranges

$$4 \times 17,250W \times .80 = 55,200W \text{ (largest two kitchen loads smaller)}$$
[NEC 220.56 and Table 220.56]

$$55,200W \ / \ 240V \times 1.732 = \underline{132.79A}$$

TOTAL - motors + other loads - 189.5A + 73A + 48.96A + 132.79A = 444.25A

According to Table 310.15(B)(16), a set of 800 kcmil THHN aluminum conductors can supply both motor and non-motor loads based on the calculated ampacity.

However, because the temperature ratings of the terminating means for the feeder conductors are unknown, the ampacity of the feeder conductors must be determined based on 75°C for circuits rated more than 100A, according to NEC 110.14(C)(1)(b).

Although an aluminum conductor having type THHN insulation can be used, the ampacity of such conductor must be limited to the ampacity listed for 75°C.

As a result, a set of 1000 kcmil THHN aluminum conductors (rated ampacity, 500A @90°C) must be used instead where the ampacity of the conductors is limited to 445A (75°C) which will suffice the calculated ampacity of 444.25A.

The results of the above calculations were based on the summation of all motors and other loads. With that being said, let's now evaluate the 3-phase feeder per-line to determine each line load. Afterward, the size of the feeder will be determined based upon the largest line load.

	L_1	L_2	L_3	N
Motors				
3HP 1φ	-	17A	17A	-
7.5HP 1φ	-	40A	40A	-
10HP 1φ (@ 1.25)*	**62.5A**	**62.5A**	-	-
10HP 3φ	28A	28A	28A	-
15HP 3φ (@ 1.25)**	42A	42A	**52.5A**	-
Lighting	36.5A	-	36.5A	36.5A
Receptacles	24.48A	-	24.48A	24.48A
Kitchen Equip.	<u>132.79A</u>	<u>132.79A</u>	<u>132.79A</u>	<u>-</u>
	326.27A	322.29A	**331.27A**	60.98A

* Largest Motor - (L_1 and L_2) ** Largest Motor - (L_3)

As you can see, line L_3 provides the largest line load. Therefore, the feeder conductors will be sized accordingly. Observe how the single phase motor loads are connected per line in an attempt to balance the loads between each line. Notice how none of the line loads contain all loads. The lighting and receptacle neutral loads had to be connected exclusively between lines L_1 and L_3 because line L_2 is the designated high-leg which results to a line to neutral voltage being other than 120V (208V). If this layout was given to an electrician, the loads would have to be connected "as shown" to maintain such balance, otherwise a given line conductor could be overloaded where the rated ampacity of the conductor is exceeded.

Based on the results of line L₃, **331.27A**, as a minimum, a set of 600 kcmil THHN aluminum conductors (rated ampacity, 385A) must be used where the ampacity at 75°C is 340A.

Just remember however, that the loads has to be connected accordingly (the way they were arranged and calculated) to eliminate adverse imbalance and overloading of the feeder conductors.

For additional questions pertaining to continuous and noncontinuous feeder loads see questions of Article 215 (Volume 2).

28. What size copper feeder conductors are needed to supply the loads shown in Figure 430.24-28. Assume the temperature rating of all terminating means is 60°C.

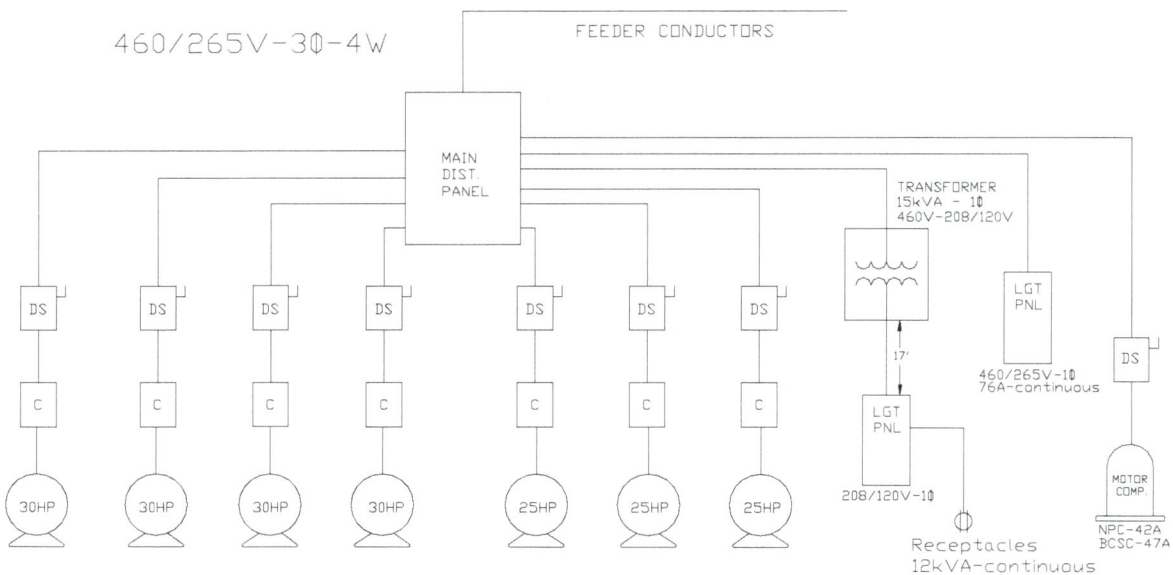

Figure 430.24-28

Referring to Table 430.250 the following full-load currents are provided:

Horsepower Rating	Full-load current
25 (3φ)	34A
30 (3φ)	40A

Referring to NEC 440.33 the ampacity of the motor-compressor is based on the largest of either the nameplate current (NPC) or the branch-circuit selection current (BCSC) marked on the motor-compressor nameplate. Because the branch-circuit selection current provides the largest current rating, this current value is used.

Motor-compressor ampacity = 47A

NEC 430.24 and 440.33 are both similar in regards to the 125 percent increase factor, however, NEC 440.33 exclusively states, "plus 25 percent of the highest motor or motor-compressor rating in the group". In this question the motor-compressor is the highest rated motor, see NEC 440.7.

$$47A \times 1.25 = 58.75A + (40A \times 4) + (34A \times 3) = \underline{320.75A}$$

Considering all other loads,

Lighting Panelboard-Receptacles (continuous) - 12kVA receptacle load given

[NEC 215.2(A)(1)] - Receptacles rated for continuous duty, no derating
12kVA / 460V x 1.732 = 15.1A (Primary load)
15.1A x 1.25 = 18.88A

Lighting Panelboard (continuous) - 76A lighting load given

[NEC 215.2(A)(1)] - 76A x 1.25 = 95A

TOTAL - motors + other loads

320.75A + 18.88A + 95A = 434.63A

According to Table 310.15(B)(16), if a set of single conductors are used as the feeder conductors, 900 kcmil copper conductors rated for 435A at 60°C must be used. However, if a set of parallel conductors are desired to serve as the feeder conductors, 1/0 AWG conductors are the smallest size conductors that can be used and the applicable factor(s) must be applied. See NEC 310.10(H), 310.15(B)(3), 310.15(B)(5) and the corrections factors, if applicable.

See question No. 26. of Article 240 (Volume 2) and question No. 8. of Article 450.

430.24, *Exception No. 1* - Several Motors or a Motor(s) and Other Load(s)

29. The following group of 3-phase motors will be fed from a common 440-volt feeder:

HP	Duty Cycle (Rated Time)	Namplate Current
20	Short-time (15 min.)	21A
25	Varying (continuous)	28A
30	Continuous	35A
50	Periodic (30 and 60 min)	58A
60	Intermittent (5 min.)	69A

Applying each motor's duty cycle and nameplate current, determine the size THWN copper conductors needed to supply the group of motors. The conductors will terminate in equipment rated for 75°C.

After realizing in question No. 22., that the operating current for motors having duty-cycles other than continuous is not determined by Table 430.250 but is based on the motor's nameplate current and finally derived by the percentage values provided in Table 430.22(E), the same must be done to established the operating currents for the motors in this question.

In compliance with NEC 430.24, *Exception No. 1* - the highest rated motor of the group must be determined as outlined. Just remember, the nameplate current rating is used for motors having a duty cycle other than continuous duty and the full-load current is used for motors rated for continuous duty. Referring to Table 430.22(E),

20HP [Short-time (15 min.)]   21A x 1.20 = 25.2A
25HP [Varying (continuous)]   28A x 2.00 = 56A
30HP [Continuous] (full-load current)*   = 40A
50HP [Periodic (30 and 60 min.)]   58A x .95 = 55.1A
60HP [Intermittent (5 min.)]   69A x .85 = 58.65A (Highest Rated)

*Table 430.250

As determined, the challenge was to identify the highest rated motor according to NEC 430.24, *Exception No. 1*. After doing so all remaining motors can now be added together to include the 30HP continuous duty motor at 100 percent of its full-load current only.

$$58.65A \times 1.25 + (25.2A + 56A + 40A + 55.1A) = 249.61A$$

Based on an ampacity of 249.61A, 250 kcmil THWN copper conductors rated for 75°C are required according to Table 310.15(B)(16).

430.25 - Multimotor and Combination-Load Equipment

NEC 430.25 reference special equipment that consists of motor(s) and other combined loads such as lighting, appliances, heating, etc. Where factory-wired, the nameplate of such equipment should identify the minimum circuit ampacity thus eliminating the need to perform load calculations to determine the proper size feeder conductors needed. Where such equipment does not come factory-wired, NEC 430.25 states that the feeder conductors must be determined based on NEC 430.24.

430.28 - Feeder Taps (See question No. 43.) NEC 430.53(D) is referenced for single motor taps.

Table 430.29 - Conductor Rating Factors for Power Resistors (Constant Voltage Direct-Current Motors-Power Resistors)

30. A 20 horsepower, 240V DC motor is being used to drive a certain load via a motor controller where power accelerating and dynamic braking resistors are used in the motor's armature circuit. The motor will operate on a 15 seconds on - 45 seconds off duty cycle. Determine the ampacity of the conductors that will be run from the controller to the resistors? If an armature shunt resistor rated for 17.3 amps is also used, what must the ampacity of the

conductors now be? Determine the ampacity of the conductors that's required to supply the armature shunt resistor.

For such installation, NEC 430.29 requires the ampacity of conductors connecting the motor controller to separately mounted power accelerating and the dynamic braking resistors in the armature circuit to have an ampacity not less than the value calculated from Table 430.29 using the motor's full-load current.

Referring to Table 430.247 the full-load current of the given motor is 72 amps. Based on the motor's duty cycle, the ampacity of the conductors per Table 430.29 must be calculated at 65 percent (.65) of the motor's full-load current, therefore,

$$72A \times .65 = 46.8A$$

The ampacity of the conductors that will be run from the controller to the resistors must be no less than 46.8 amps.

If an armature shunt resistor is used, the ampacity of the conductors according to NEC 430.29 must be increased by the 17.3 amps current rating of the resistor thus requiring the ampacity of the conductors be no less than 64.1 amps (46.8A + 17.3A).

In concluding, NEC 430.29 also requires the ampacity of the armature shunt resistor conductors to have an ampacity not less than that calculated from Table 430.29 using the rated shunt resistor current as the full-load current. As a result, 17.3A x .65 = 11.25A.

The ampacity of the armature shunt conductors must be no less than 11.25 amps.

SIZING MOTOR CIRCUIT OVERLOAD PROTECTION

An overload protective device must protect and de-energize motors, motor-compressor equipment, motor-control circuitry and motor circuit conductors in the event of an overload which includes locked-rotor current. In accordance with NEC 430.6(A)(2), when determining the setting of a motor's overload protection device, the actual current rating marked on the motor's nameplate is used instead of the full-current values of Tables 430.247 through 430.250. The *service factor* and *temperature rise* of a motor are other factors used for determining the size of a motor's overload protection device. A motor's service factor is the allowable horsepower overload the motor is capable of sustaining. Service factor ratings normally range between 1 and 1.25. The temperature rise of a motor simply indicates the difference between the operating temperature of a motor and the surrounding temperature the motor could possibly be exposed to. Motor manufacturers usually design motors to operate at no less than 40°C (104°F) above the surrounding temperature. When a motor is placed in an environment where the surrounding temperature is greater than the motor's operating temperature, the heat produced by the motor cannot dissipate (move away, spread out), making it extremely difficult for the motor to breathe. Bearing friction and eddy currents also play an important part towards the contribution of heat. Typically, heat is the leading cause of most motor failures.

ACCORDING TO THE NEC

The following NEC sections are applicable for understanding how to size motor overload protection per Part III of Article 430.

NEC 430.6(A)(2) - Nameplate Values - Separate motor overload protection shall be based on the actual nameplate current rating of a motor.

NEC 430.32(A)(1) - Separate Overload Device (More Than 1 Horsepower [Continuous-Duty Motors]) - This section pertains to continuous duty motors rated for more than 1 horsepower. A separate overload device that is responsive to motor current must be selected to trip or be rated no more than the following percent (minimum) of the motor's nameplate full-load current rating:

Motors with a marked service factor (SF) of 1.15 or greater	125% (1.25)
Motors with a marked temperature rise (TR) of 40°C or less	125% (1.25)
All other motors	115% (1.15)

For multispeed motors, each winding connection must be considered separately. If a separate overload device is separately connected and fails to carry the total current specified on the motor nameplate, such as for wye-delta starting, the proper percentage of the appropriate current for either the selection or the setting of the overload device must be clearly marked on the equipment involved or gathered from the manufacturer's selection table.

NEC 430.32(B)(1) - Separate Overload Device (One Horsepower or Less, Automatically Started [Continuous-Duty Motors]) - This section pertains to continuous duty motors rated less than 1 horsepower that are automatically started. The requirements for sizing separate overload devices are the same as NEC 430.32(A)(1).

NEC 430.32(B)(2) - Thermal Protector (One Horsepower or Less, Automatically Started [Continuous-Duty Motors]) - This section is referenced to determine the ultimate trip current for motors that are thermally protected. Thermal protectors are used to sense excessive temperature or current.

NEC 430.32(C) - Selection of Overload Relay (Continuous-Duty Motors) - When the percentage values of NEC 430.32(A)(1) is not sufficient, this section allows higher percentage (maximum) values of a motor's nameplate full-load current rating for sizing overload protection based on the following:

Motors with a marked service factor (SF) of 1.15 or greater	140% (1.40)
Motor with a marked temperature rise (TR) of 40°C or less	140% (1.40)
All other motors	130% (1.30)

NEC 430.36 - Fuses - In Which Conductor - Where fuses are used for motor overload protection, a fuse must be inserted in each ungrounded conductor and also in the grounded conductor if the supply system is 3-wire, 3-phase ac with one conductor grounded.

NEC 430.37 - Devices Other Than Fuses - In Which Conductor - Where devices other than fuses are used for motor overload protection, Table 430.37 shall govern the minimum allowable number and location of overload units such as trip coils or relays.

APPLYING THE NEC

According to Article 100 the term overload is defined as the operation of equipment in excess of normal, full-load rating, or of a conductor in excess of rated ampacity that, when it persists for a sufficient length of time, would cause damage or dangerous overheating. In concluding, the definition also states that a fault, such as a short circuit or ground fault, is not an overload.

Part III of Article 430 governs the use of overload protection (commonly referred to as a heater, an integral part of a motor starter) for protecting motors, motor-control equipment and branch-circuit conductors against excessive heating due to motor overloads and failure to start. According to NEC 430.32(A) motors having a continuous duty cycle and rated for more than 1 horsepower must be provided with some means of overload protection. Thermal protectors, thermal relays and fuses* are the most common means of overload protection referenced in this section. NEC 430.32(A)(1) provides procedures for sizing separate overload devices which concur with the operating conditions of the motor. Motors that are marked with a service factor not less than 1.15 *or* has a temperature rise not exceeding 40°C requires the nameplate current of the motor to be increased by 125 percent (1.25). When the service factor and temperature rise of a motor fail to meet given requirements *or* are not provided; the nameplate current of the motor must only be increased by 115 percent (1.15).

*NEC 430.36 and 430.55

service factor - The permissible overload that a motor is designed to withstand. A motor's service factor is determined by multiplying the motor's assigned service factor value by the horsepower (HP) rating of the motor. For example, if a 25HP motor's assigned service factor is 1.15 then the permitted overload or horsepower allows the motor to operate at 28.75HP (25HP x 1.15). To operate at its assigned service factor the motor's rated voltage and frequency must be maintained. In the event a motor is caused to operate at its assigned service factor the motor's rated current and temperature rise per nameplate values will be exceeded.

temperature rise - The difference between the operating temperature of a motor and the ambient (surrounding) temperature. The temperature a motor is designed to operate above the ambient temperature. For example, if a motor rated for continuous duty has a 30°C temperature rise, the motor will stabilize its temperature at 30°C above the ambient temperature without the rise in temperature causing damage to the windings and other vital parts of the motor. Heat is a major contributor to motor failure.

When sizing the overload protective device for motors, always use the motor's nameplate current as shown on the nameplate of the motor, if available (otherwise apply the values given in Tables 430.247 through 430.250) per application. Apply NEC 430.32(A)(1) initially to satisfy the minimum requirements for sizing overload protection. If the minimum sized overload protection causes nuisance tripping or proves to be insufficient to start the motor, NEC 430.32(C) can then be applied. Refer to the following procedures,

1. Determine the motor's nameplate current as marked on the motor's nameplate.

2. Apply NEC 430.32(A)(1) to determine the *minimum* size overload protection for motors rated for continuous duty.

3. When insufficiently sized, apply NEC 430.32(C) to determine the *maximum* size overload protection for motors rated for continuous duty.

430.32(A)(1) - Separate Overload Devices (More Than 1 Horsepower) (31. - 32.) [2]

31. Using the nameplate data of a dual-rated motor where the motor's nameplate currents are 170A at 230V and 85A at 460V, determine the minimum size overload protective devices required to protect the motor. The motor has a 30°C temperature rise and 1.15 service factor.

Question Nos. 31., 32. and 34. will be used as examples to demonstrate the procedure for sizing motor overload protection.

Application

Step 1: Determine the motor's nameplate current as marked on the motor's nameplate.

Because the motor is dual rated, it list two nameplate currents, 170A at 230V and 85A at 460V.

Step 2: Apply NEC 430.32(A)(1) to determine the minimum size overload protection for motors rated for continuous duty.

Because the motor has a service factor and temperature rise which results in using the same percentage increase of the motor's nameplate current, the minimum size overload protective devices to be used per voltage and nameplate current ratings are:

at 230V - 170A x 1.25 = 212.5A   at 460V - 85A x 1.25 = 106.25A

If the motor is used for 230 volts the minimum size overload protective devices must be rated for 212.5A and if used for 460 volts the minimum size overload protective devices must be rated for 106.25A.

See question No. 21. of Article 440 for motor-compressor overload protection.

32. Refer to Figure 430.32(A)(1)-32. Determine the minimum size overload protection needed for each motor. This question is a continuation of question No. 25.

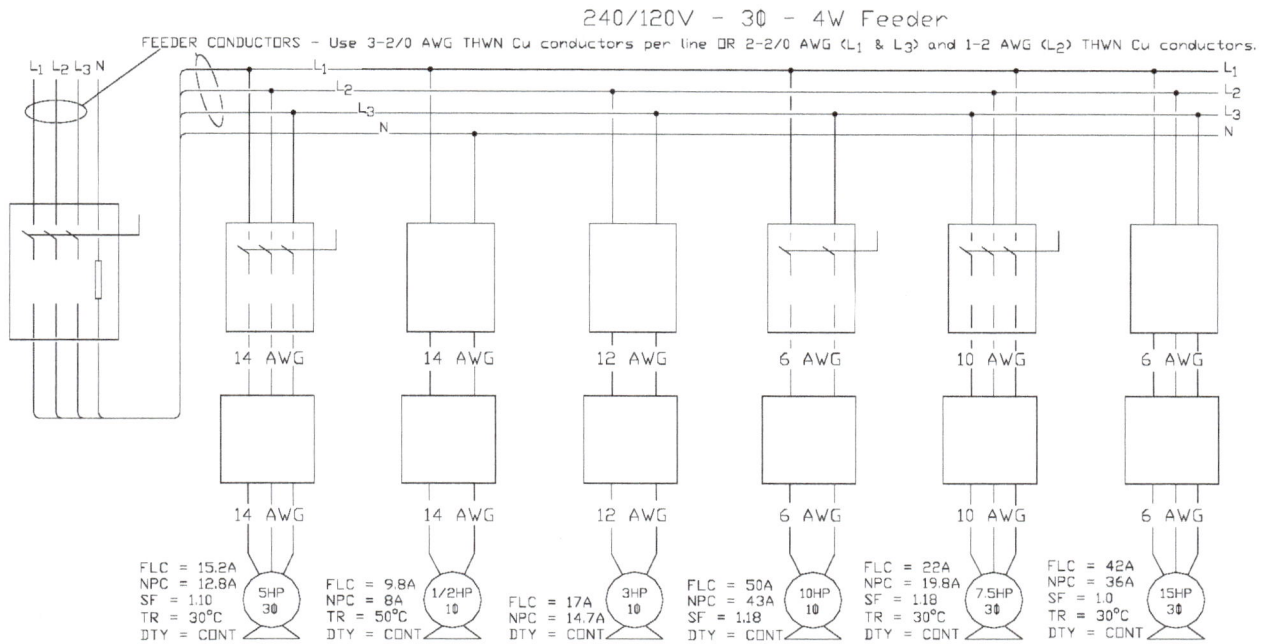

Figure 430.32(A)(1)-32

Application

Step 1: Determine the motor's nameplate current as marked on the motor's nameplate.

Refer to the data provided for each motor in Figure 430.32(A)(1)-32.

Step 2: With the exception of the ½ horsepower motor, NEC 430.32(A)(1) can be applied to determine the minimum size overload protection for motors over 1 horsepower and rated for continuous duty. Although the same percentage values are applied, NEC 430.32(B)(1) is referenced to determine the minimum size overload protection for the ½ horsepower motor.

Using each motor's nameplate current and other needed data provided in Figure 430.32(A)(1)-31, the minimum overload protection is determined for each motor.

$$5HP - 12.8A \ x \ 1.25 = 16A$$
$$½HP - \ 8.0A \ x \ 1.15 = \ 9.2A$$
$$3HP - 14.7A \ x \ 1.15 = 16.91A \quad \text{No data, classified as "other motor".}$$
$$10HP - 43.0A \ x \ 1.25 = 53.75A$$
$$7.5HP - 19.8A \ x \ 1.25 = 24.75A$$
$$15HP - 36.0A \ x \ 1.25 = 45A$$

430.32(A)(2) - Thermal Protector (More Than 1 Horsepower)

33. If a 3-phase, 460V, 75HP motor was equipped with a thermal protector, determine the ultimate trip current setting of the protective device.

Per NEC 430.32(A)(2), where a motor's full-load current is greater than 20 amps the ultimate trip current setting for a thermal protector must be calculated at 140 percent (1.40) of the affiliate motor's full-load current. The full-load current for a 3-phase, 460V, 75 horsepower motor per Table 430.250 is 96 amps. As a result, the thermal protector's ultimate trip current must not exceed 134.4 amps.

$$96A \times 1.40 = 134.4A$$

430.32(B)(1) - Separate Overload Device (One Horsepower or Less, Automatically Started)

34. A ½HP single-phase motor rated for 208V has a nameplate current marked for 3.7A. The motor has a 1.10 service factor and a 30°C temperature rise. The motor is automatically started and runs no less than 2 hours when operated. What size overload protective device is required to protect the motor? If the motor experience nuisance tripping due to undersized overloads, determine the correct size overloads needed.

Application

Step 1: Determine the motor's nameplate current as marked on the motor's nameplate.

The motor's marked nameplate current at 208V is 3.7A.

Step 2: Apply NEC 430.32(A)(1) to determine the minimum size overload protection for motors rated for continuous duty.

Because the percentage values found in NEC 430.32(A)(1) are the same as required in NEC 430.32(B)(1), motors rated less than a horsepower that operates continuously when automatically started use the same guidelines for sizing overload protection per NEC 430.32(A)(1).

Although the motor's service factor is less than 1.15, the temperature rise is not over 40°C. Again, just remember that the requirements of NEC 430.32(A)(1) are not based on both conditions (when provided) being met. Therefore, 3.7A x 1.25 = 4.63A.

Step 3: When insufficiently sized, apply NEC 430.32(C) to determine the maximum size overload protection for motors rated for continuous duty.

According to NEC 430.32(C), the percentage value of the motor's nameplate current can be increased to 140 percent (1.40) based on the above conditions. This value can be used when the overload protection sized according to NEC

430.32(B)(1) proves to be insufficient or produces nuisance tripping of the motor due to undersized overload protection. As a result, 3.7A x 1.40 = 5.18A.

If the motor's overload protective device produces nuisance tripping at 4.63A the rating of the overload protective devices can be increased to 5.18A.

430.32(C) - Selection of Overload Relay

35. If the overload protection selected for the motors in question No. 32. causes nuisance tripping, determine the maximum size overload protection that can be used.

Again, NEC 430.32(C) permits the percentage value of each motor's nameplate current to be increased based on the data provided in Figure 430.32(A)(1)-32. These values are permitted to be used when the overload protection sized according to NEC 430.32(A)(1) proves to be insufficient or produces nuisance tripping of the motor due to undersized overload protection. As a result,

		Per NEC 430.32(C) (maximum)	Compared to NEC 430.32(A)(1) (minimum)
5HP - 12.8A x 1.40	=	17.92A	16A
½HP - 8.0A x 1.30	=	10.4A	9.2A
3HP - 14.7A x 1.30	=	19.11A	16.91A
10HP - 43.0A x 1.40	=	60.2A	53.75A
7.5HP - 19.8A x 1.40	=	27.72A	24.75A
15HP - 36.0A x 1.40	=	50.4A	45A

430.36 - Fuses – In Which Conductor

36. If time-delay, dual element fuses are used to serve as the overload protection for a three-phase, 15HP motor rated for 230V, what size fuses are required? The motor has a 1.18 service factor and nameplate current marked for 35A.

Application

Step 1: Determine the motor's nameplate current as marked on the motor's nameplate.

The motors marked nameplate current at 230V is 35A.

Step 2: Apply NEC 430.32(A)(1) to determine the minimum sized overload protection for motors rated for continuous duty.

The use of fuses as overload protection for a motor is permitted in NEC 430.36. When fuses are used they must be inserted between each ungrounded conductor to include the grounded conductor if the source of power is derived from a 3-wire, 3-phase *ac* system with one conductor grounded, such as a 3-wire, 3-phase corner-grounded delta system. The voltage rating for this type system usually ranges from 230V and higher.

NEC 240.22 requires the exclusion of an overcurrent device for a conductor that is intentionally grounded unless adhering to the provisions of NEC 430.36 or 430.37. A copper "slug" or "busbar" is usually inserted in series with the conductor that's grounded.

Using the given nameplate current and service factor based on NEC 430.32(A)(1) yields the following, 35A x 1.25 = 43.75A. Rounding the fuse up to the next standard size would be in violation of NEC 430.32(A)(1) because it would exceed the125 percent limitation, therefore 40A time-delay, dual element fuses are required to serve as the motors overload protection.

Step 3: When insufficiently sized, apply NEC 430.32(C) to determine the maximum size overload protection for motors rated for continuous duty.

If the 40A fuses prove to be inadequate, then NEC 430.32(C) can be applied, where

$$35A \times 1.40 = 49A$$

Once again, the fuse size cannot be rounded up to the next standard size because it would then be in violation of NEC 430.32(C). As a result, the use of 45A time-delay, dual element fuses are allowed.

Since fuses are normally used to provide overcurrent (short-circuit and ground-fault) protection for motors when used for overload protection they serve a dual purpose. Because the percentage values (Table 430.52) used to size overcurrent protection for motors are not as stringent as those used for sizing overload protection, the selected fuses will respond to overcurrent (due to overload) much sooner.

In addition, just remember that in sizing the overcurrent protection for motors; the motor's full-load current is used, resulting in a higher current rating than the nameplate current of a motor which is used for sizing overload protection.

SIZING MOTOR CIRCUIT OVERCURRENT PROTECTION

One of the most misleading assumptions about an overcurrent device, be it fuse or circuit breaker is, the device will prevent short circuits or equipment failure. Considering the immeasurable rate of speed of a short-circuit just imagine a device that could totally interrupt such a destructive occurrence.

Table 430.52 is used to establish the maximum rating or setting of a motor's branch-circuit overcurrent device. NEC 430.6(A)(1) states that the full-load current values provided in Tables 430.247 - 430.250, must be used for determining the ampere rating of a motor's branch-circuit overcurrent device, instead of the actual current rating marked on the motor's nameplate. Concurring with NEC 430.52(C)(1), the contents of Table 430.52 provides a listing of percent multipliers for increasing a motor's full-load current. The first column of the table identifies the various types of motors that are subject to the provisions of Table 430.52. The remaining four columns list a combination of overcurrent devices consisting of fuses and circuit breakers. Each set of percent multipliers per column represents the value a motor's full-load current is increased with regards to determining the rating or setting of the required overcurrent device. When the calculated values for branch-circuit overcurrent devices, as derived from Table 430.52, does not correspond to a standard size overcurrent device per NEC 240.6(A) the next higher standard size overcurrent device is allowed. Where the selected size of a protection device is insufficient for the starting current of a motor NEC 430.52(C)(1), *Exception No. 2* and 430.52(C)(3), *Exception No. 1* permits the use of a larger size overcurrent device.

BRANCH-CIRCUIT OVERCURRENT PROTECTION

<u>ACCORDING TO THE NEC</u>

The following NEC sections and tables are applicable for understanding how to size motor branch-circuit short-circuit and ground-fault protection devices per Part IV of Article 430.

NEC 240.4(D) - Small Conductors - The provisions of this section does not apply to motor overcurrent branch-circuit protection devices. Refer to NEC 240.4(G).

NEC 240.6(A) - Fuses and Fixed-Trip Circuit Breakers (Standard Ampere Ratings) - The standard ampere ratings for fuses and inverse time breakers are as follows,

---

15, 20, 25, 30, 35, 45 – 50, 60, 70, 80, 90, 100, 110 – **125, 150, 175, 200, 225, 250**

300, 350, 400, 450, 500 – **600, 700, 800, 1000** – 1200, 1600, 2000 – **2500**

3000, 4000, 5000, 6000

Additional standard ratings for fuses shall be 1, 3, 6, 10 and 601 amperes. The use of fuses and inverse time circuit breakers with nonstandard ampere ratings is allowed.

NEC 430.6(A)(1) - Table Values - When sizing motor circuit conductors the full-load current values provided in Tables 430.247 - 430.250, including related notes, shall be used to determine the ampacity of conductors (motor circuit) or ampere ratings of switches, branch-circuit short-circuit and ground-fault protection (circuit breakers and fuses) instead of the actual current rating marked on the motor nameplate.

Table 430.52 - Maximum Rating or Setting of Motor Branch-Circuit Short-Circuit and Ground-Fault Protective Devices - Table 430.52 is used for sizing the overcurrent protection device for motors. The percentage values listed in the table for the four given overcurrent protection devices usually allows a motor to start above its locked-rotor current and reach its operating speed without tripping or opening the device. Particular notice should be given to the footnotes listed below the table. Where the percentage values of the table provides inadequate ampere ratings or prove to be insufficient, refer to *Exception Nos. 1* and *2* to NEC 430.52(C)(1) and/or NEC 430.52(C)(3).

UNDERSTANDING TABLE 430.52

COLUMN 1 - **Type Of Motor**

Column 1 list **7** types of motors that rely upon the percentage values provided in **Table 430.52** for selecting the appropriate branch-circuit overcurrent protection device for each individual motor type. This column reference single-phase motors, two and three-phase *ac* motors, squirrel cage, design *B* energy efficient motors, synchronous motors, wound-rotor motors and direct-current *(dc)* motors. Before proceeding lets identify each motor type in terms of physical structure and operating characteristics.

SINGLE-PHASE MOTORS

Although Table 430.248 limits single-phase motors to 10 horsepower, the horsepower ratings of such motors can go much higher. Most single-phase motors are designed for dual voltage operation, meaning they can either be run on a minimum of 115V or 230V (approximately). Unlike three-phase motors, single-phase motors must have supplemental means for starting the movement of its rotor. Once rotor movement occurs, the rotor induces a rotating magnetic field about the motor that sustains the operation of the motor. For such reason, the starting circuit of a single-phase motor is either opened or remains closed after the rotor reaches a predetermined speed. When selecting a single-phase motor, the deciding factor will always be based upon operating needs, starting application and torque.

Types of single-phase motors

Split-Phase - Commonly called an induction-start/induction-run motor, split-phase motors are one of the least expensive/less complicated motors of all single-phase motors. Usually ranging up to 1/3 horsepower, split-phase motors are mostly used for tool grinders, small fans, blowers and other applications requiring low-starting torques. Split-phases motors are equipped with two types of windings, a start winding and a run (main) winding. The start winding is formed with wire smaller than the run winding, thus resulting in a much higher resistance. This causes the current flow in both windings to differ, which in terms offset the intensity of each winding's magnetic field. With two magnetic fields, where one is displaced about 30° from the other, a rotating field is formed which causes the rotor to turn. Once a split-phase motor reaches about 75 percent of its rated speed the start winding is discontinued by a centrifugal switch while the run winding remains in operation.

Permanent-Split Capacitor - The permanent-split capacitor motor is very similar to a split-phase motor, where the difference is a run-type capacitor is installed in series with the start winding, which remains functional, while the motor is in full operation. This motor is considered to be the most reliable among other single-phase motors, mainly because a starting switch is not required. Made in sizes up to 3 horsepower, permanent-split capacitor motors are built for low starting torque applications such as for heating and air conditioning blowers and where reversing and intermittent cycling is needed.

Shade-Pole - Unlike other single-phase motors, shade-pole motors have only one run winding and no start winding. Starting a shade-pole motor requires equipping a small portion of each stator pole with a supplemental winding called a shaded coil. Although, the shaded coil is electrically isolated, it depends upon induced current to create a rotating magnetic field to turn the rotor. More suitable for household appliances with fans and blowers where motor requirements are not more than ¾ horsepower, these motors have low starting torque and are limited to light-duty applications.

Capacitor-Start, Induction-Run - Similar in features to a split-phase motor, capacitor-start, induction-run motors have a high starting torque and are often used for heavy loads such as air compressors, elevators and other high demand applications. These motors are usually available in sizes ranging from a fractional horsepower up to 10 horsepower. Connected in series with the start winding and a centrifugal switch is a capacitor that is mounted either on top or to the side of the motor. Once the motor reaches about 75 percent of its operating speed the start winding and capacitor are de-energized by the centrifugal switch.

Capacitor-Start, Capacitor-Run - Where the need exist for demanding applications, this single-phase motor provides great results. Capacitor-start, capacitor-run motors are used for driving hard starting loads such as air compressors, high-pressure water pumps, vacuum pumps and other high-torque applications. These motors are usually available in sizes ranging from ½ horsepower up to 25 horsepower. Two capacitors are used in a capacitor-start, capacitor-run motor. One capacitor has a higher capacitance for starting the motor while the other has a lower capacitance for running the motor. Because both capacitors are connected in parallel the total capacitance between both capacitors is added together when the motor is started, thus resulting in a high starting torque. After the motor is started and reaches running speed, a centrifugal switch discontinues the use of the starting capacitor leaving the running capacitor energized to provide the motor with a higher running torque.

Series - Often called a universal motor, series motors are used to operate small equipment such as portable drills, power saws, food mixers and vacuum cleaners. A series motor can operate on either *ac* or *dc* voltages and are mostly built in sizes of 1 horsepower or less. The speed of a series motor is not constant and varies according to the applied load. For heavy loads, the speed may get up to a few hundred rpms compared to a speed of 15,000 rpms for no-load operations. Because series motors are not induction type motors they are design differently. Series motors consist of field windings and an armature with brushes and a commutator. The commutator is made of insulated segments, each of which is connected to an armature winding. Brushes set on the commutator conducts current from one field winding through the armature winding via the commutator to an opposite field winding. The commutator keeps the armature turning through the magnetic field produced by the field windings. Once the flow of current changes direction in the field windings and the armature, the magnetic field is reversed thus causing the armature to continue rotating.

Repulsion-Start, Induction-Run - For loads that are difficult to start, repulsion-start, induction-run motors are made exclusively for that purpose. These motors come in sizes ranging from 1/6 horsepower to 20 horsepower. They are mostly used for water pumping and rock crushing applications. Having an armature, a commutator and brushes make repulsion-start, induction-run motors quite similar to series motors. In a repulsion-start, induction-run motor, the field and armature windings are not connected in series and the brushes are linked together. Because the armature windings are not connected to the field windings serially, the current flowing through the armature windings is induced. Upon approaching full speed, a centrifugal switch lifts the brushes and short-circuits the commutator segments with a short-circuit ring or short-circuiter. Once the brushes are lifted and the commutator shorted out, the armature becomes a rotor and the motor operates as an induction motor.

POLYPHASE MOTORS

The word *polyphase* means having more than one phase, therefore the *ac* polyphase motors identified in **Table 430.52** reference two and three-phase squirrel cage motors rated for standard or energy efficient use.

Types of polyphase motors

Two-phase - In some parts of the United States two-phase motors are still being used. Needless to say however, they are being phased out and replaced with three-phase motors. To familiarize those who are unfamiliar with this type motor, the windings of a two-phase motor are placed 90° apart thus making the voltage relationship between the windings 90° out of phase with each other. Table 430.249 of Article 430 list the full-load currents of a two-phase motor connected to a 4-wire system. As shown in Figure 430.249-1 a 4-wire, two-phase system consists of two separate single-phase circuits, which are electrically isolated from each other.

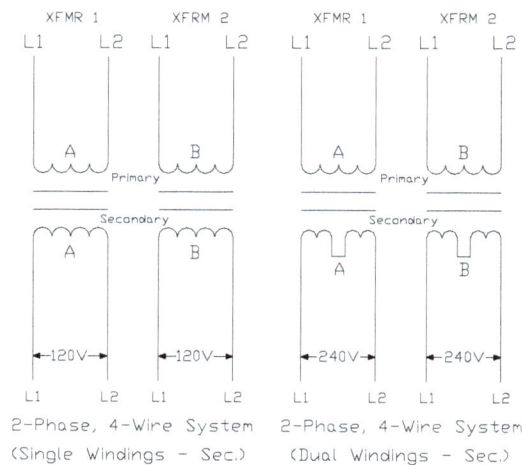

Figure 430.249-1

Figure 430.249-2 displays a 3-wire, two-phase system. The voltage between the two phases of a 3-wire system is the product of multiplying the single-phase voltage by the square root of two ($\sqrt{2}$) or simply 1.414.

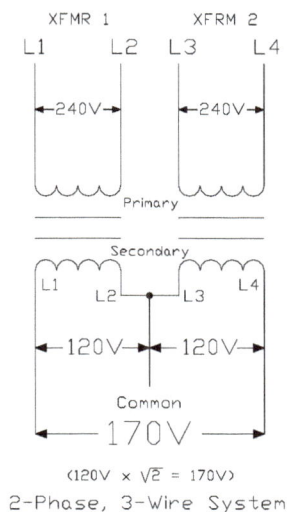

Figure 430.249-2

As referenced in the statement following the heading of Table 430.249, current in the common conductor of a 3-wire, 2-phase system will be 1.41 times the applied full-load current value given in the table. In general, the principles of operation for a two-phase motor are identical to that of a three-phase motor.

Three-phase - By far, three-phase motors are the most commonly used motors in the United States and abroad. They are manufactured in sizes ranging from fractional to several thousand horsepower, low in cost and generally require less maintenance and repair than single-phase motors. Unlike single-phase motors, most three-phase motors are self-starting and capable of starting heavy loads. They don't require a start winding, capacitors, centrifugal switches or other starting mechanisms. Three-phase motors are operated by a three-phase power source where the voltage between each phase is 120° out of phase with each other. The three basic types of three-phase motors are the squirrel cage, synchronous and wound rotor motors. Squirrel cage and wound rotor motors rely on the principles of induction and rotating magnetic fields for starting and operating.

SQUIRREL CAGE (other than Design B energy efficient)

The squirrel cage induction motor of all three-phase motor types is the motor most often utilized. As mentioned, this self-starting motor only relies upon the interaction of its stator and rotor magnetic fields based on the principles of induction to operate. Unlike a wound-rotor motor it requires neither brushes nor slip rings thus increasing the durability and roughness of the motor. The squirrel cage motor can be connected for either high or low voltage if rated for dual-voltage.

DESIGN B ENERGY-EFFICIENT

The ratio of mechanical power output to the electrical power input defines the efficiency of a motor or motor efficiency. Such results are typically displayed in a percentage value less than 100 percent. Design improvements, better manufacturing methods along with the use of materials of higher quality enables energy-efficient motors to exceed the performance of a standard motor by yielding a higher work output at a lower consumption of energy.

Because the construction of an energy-efficient motor is superior to that of a standard motor, an energy-efficient motor usually offers longer insulation and bearing life, higher service factors, less vibration, and lower waste heat output; all of which contributes to an increase of motor reliability. Based on the reliability of this type motor, most motor manufacturers offer longer manufacturer's warranties to that of a standard motor.

To be considered "energy-efficient", a motor must perform at a level that is either equal to or exceed the nominal full-load efficiency values provided by the National Electrical Manufacturers Association (NEMA) standard MG-1. This publication provides specific full-load nominal efficiency values for each horsepower rating, enclosure type, and speed combination.

The Energy Policy Act of 1992 (EPACT) requires most general purpose motors manufactured for sale in the United States to meet NEMA standards. This Act applies to motors with horsepower ratings ranging from 1 to 200-horsepower that conforms to NEMA designs A and B.

SYNCHRONOUS MOTORS

Three-phase synchronous motors are highly efficient and are mostly recognized for their ability to produce a leading current. Normally the efficiency of a synchronous motor range between plus or minus 5 percent of a 90 percent rating. A synchronous motor can improve the power factor of an electrical system supplying induction motors due to the leading current produced by a synchronous motor and the lagging current produced by an induction motor. Because both currents opposes each, virtually canceling each other out, the supplying electrical system is required to produce very little reactive current, which in terms reduces the system's demand for producing useful current (resistive), therefore improving the electrical system's power factor. Unlike squirrel cage and wound rotor motors, synchronous motors are not self-starting. They operate at a constant speed whether no-load or a full-load is applied. These motors can be purchased to meet either low or high speed demands.

WOUND ROTOR MOTORS

Three-phase wound rotor motors can provide high starting torque and low starting current coupled with smooth acceleration for high inertia and heavy duty loads. Compared to squirrel cage motors, the rotor of a wound rotor motor has insulated (wound coils) windings opposed to metal bars. Three leads from the rotor windings are connected to an external variable resistor that changes the resistance of the rotor windings. As the external resistance of a wound rotor motor increases, current flow in the rotor windings decreases and vice-versa causing the speed of the motor to change likewise. Although wound-rotor motors are classified as variable-speed motors, the speed range is limited and for such reasons wound rotor motors are not often used for variable speed applications.

DIRECT CURRENT MOTORS (constant voltage)

Compared to an alternating current (*ac*) motor, the physical characteristics of a direct current (dc) motor are quite similar but that's as far as it goes. While *ac* motors operate on alternating current, *dc* motors on the other hand obviously operates on direct current. Direct current motors come in several different types where series, shunt and compound motors are the type *dc* motors most frequently used. A *dc* motor will operate at a selected speed only with a specific combination of voltage and load. In the event the applied voltage or load changes, the speed of the motor will change also. The speed of a *dc* motor can be changed by use of a controlling device although such changes in speed are still vulnerable to any change in applied voltage or load. One of the most apparent similarities between *dc* and *ac* are both type motors employ a magnetic field to rotate their moving parts. Due to their varying speed capabilities, high starting torques and the ability to reverse the direction of rotation, *dc* motors are ideal for elevators, cranes, hoist and similar type applications. Direct current motors can be purchased in sizes ranging from very small to several hundred horsepower.

COLUMN 2 - Nontime Delay (NTD) Fuse

A **nontime delay fuse** detects short circuits and responds instantaneously when the rating of the fuse is exceeded. This type fuse operates on thermal principles and senses heat build-up in the circuit it protects. Once the element of a nontime delay fuse is blown it is no longer usable. The nontime delay fuse listed in Column 2 is just a regular type fuse and usually holds up to 500 percent (5) of its current rating for about one-quarter of a second (.25s). However, some nontime delay fuses are designed to hold up to 2 seconds (2s) at the same level. When used as the protective device for a motor, the motor must start and reach its running speed based on the operating conditions of the fuse.

COLUMN 3 - Dual Element (Time Delay Fuse)

A **dual element/time-delay fuse** detects thermal overloads and short circuit currents yet responds in a time-delayed manner when the rating of the fuse is exceeded. A dual element or time-delay fuse consist of a thermal and short-circuit element to allow the delay of a needed response over a given period of time. This type fuse will maintain a current flow up to 500 percent or five times its ampere rating for ten seconds without blowing. With such time-delay, this fuse will allow any motor to start at a current which greatly exceeds the motor's full-load current until the motor reaches its normal operating conditions.

COLUMN 4 - Instantaneous Trip (Circuit) Breaker

An **instantaneous trip circuit breaker** responds to an instantaneous value of over-current regardless of whether it's initiated by means of a short-circuit, ground fault or locked-rotor current. Instantaneous trip circuit breakers do not provide thermal protection. Some instantaneous trip circuit breakers are equipped with adjustable trip settings and will trip at approximately 300 percent or three times it's rating on low settings and at 1000 percent or 10 times it's rating on high settings. NEC 430.52(C)(3), *Exception No. 1* permits an instantaneous trip circuit breaker to be increased up to 1300 percent (13 times) of a motor's full-load current when the motor is other than a Design B energy-efficient motor and up to 1700 percent (17 times) of a motor's full-load current for Design B energy-efficient motors. An instantaneous trip circuit breaker will maintain the excessively high starting current of a motor.

COLUMN 5 - **Inverse Time (Circuit) Breaker**

An **inverse time circuit breaker** responds to both thermal overloads and short-circuits or ground fault currents. Inverse time trip circuit breakers are equipped with thermal and magnetic elements to de-energize a motor upon on the occurrence of an undesired overcurrent. The thermal element responds to heat while the magnetic element responds to short-circuit or ground fault currents exceeding the instantaneous values of the circuit breaker's rated current. Either response will cause the circuit breaker to trip. An inverse time circuit breaker works upon the principle where the sensed overcurrent determines the tripping response of the circuit breaker. In other words, the larger the over current - the shorter the tripping response time and vice-versa. Typically, an inverse time circuit breaker will maintain 300 percent or three times its rating at variable intervals ranging from 4 to 40 seconds depending upon the applied voltage and current setting of the circuit breaker.

NEC 430.52(B) - All Motors - A motor's branch-circuit overcurrent protection device must be capable of carrying a motor's starting current.

NEC 430.52(C)(1) - In Accordance with Table 430.52 (Rating or Setting) - An overcurrent device that has a rating or setting not exceeding the value calculated according to the values given in Table 430.52 must be used.

Exception No. 1 - If the calculated value for an overcurrent protection device as determined by Table 430.52 does not correspond to a standard size overcurrent protection device, as listed in NEC 240.6(A), and deemed inadequate to carry the motor's load, the next standard size overcurrent protection device is permitted.

Exception No. 2 - If the rating of the overcurrent protection device specified in Table 430.52, as modified by *Exception No. 1*, is not sufficient to carry the starting current of a motor, then:

(a) Where nontime delay or time-delay Class CC (current-limiting) fuses with an ampere rating of 600 amperes or less are used, the rating of the fuse can be increased up to no more than 400 percent (four times [4]) of the full-load current.

(b) Where time-delay fuses are used, the ampere rating of the fuse can be increased, but should never exceed 225 percent (two and a quarter times [2.25]) of the full-load current.

(c) Where inverse time circuit breakers are used, the ampere rating of the circuit-breaker can be increased up to no more than 400 percent (four times [4]) of the full-load current when the full-load current value does not exceed 100 amperes or up to no more than 300 percent (three times [3]) of the full-load current when the full-load current value exceeds 100 amperes.

(d) A fuse having a rating between 601*-6000 amperes can be increased up to no more than 300 percent (three times [3]) of the full-load current.

*A 601A fuse bolts in while a 600A fuse pushes in. A 601A fuse is approximately 2" longer than a 600A fuse. A 601A fuse serves as the allowance for a normally rated 600A fuse to be used in a higher rated fuse holder.

NEC 430.52(C)(3) - Instantaneous Trip Circuit Breaker - Instantaneous trip circuit breakers can only be used if adjustable and if part of a listed combination motor controller having coordinated motor overload and overcurrent protection in each conductor, and the setting is adjusted to no more than the value specified in Table 430.52.

Exception No. 1 - If the setting specified in Table 430.52 is not sufficient to carry the starting current of a motor, the setting of the circuit breaker shall be permitted to be increased up to no more than 1300 percent (thirteen times [13]) of the motor's full-load current for motors other than Design B energy efficient. For Design B energy efficient motors the setting of the circuit breaker shall be permitted to be increased up to no more than 1700 percent (seventeen times [17]) of the motor's full-load current. Trip settings above 800 percent (eight times [8]) for other than Design B energy efficient motors and 1100 percent (eleven times [11]) for Design B energy efficient motors are permitted once the need for such settings is verified by an engineering evaluation. When such evaluations apply, it is not necessary to initially use the 800 percent and 1100 percent values for determining the setting of an instantaneous trip circuit breaker.

Exception No. 2 - Where the motor full-load current is 8 amperes or less, the setting of the instantaneous-trip circuit breaker with a continuous current rating of 15 amperes or less in a listed combination motor controller that provides coordinated motor branch-circuit overload and overcurrent protection shall be permitted to be increased to the value marked on the controller.

NEC 430.52(D) - Torque Motors - Torque motor branch-circuits shall be protected at the motor nameplate current rating in accordance with 240.4(B).

NEC 430.53 - Several Motor or Loads on One Branch Circuit - As outlined in NEC 430.53(A), (B), (C) and (D), two or more motors or one or more motors along with other loads are permitted to be connected to the same branch circuit. The branch-circuit protective device shall be fuses or inverse time circuit breakers.

NEC 430.53(A) - Not Over 1 Horsepower - Two or more motors rated less than 1 hp may be used on a branch circuit that's rated for 120 volts and protected by an overcurrent device rated for 20 amperes or less or a branch circuit rated for 600 volts or less and protected by an overcurrent device rated for 15 amperes or less, if the full-load rating of each motor does not exceed 6 amperes and the overcurrent device marked on any of the controllers is not exceeded and the individual overload protection conforms to NEC 430.32.

NEC 430.53(B) - If Smallest Rated Motor Protected - Where determined that the branch-circuit overcurrent protection device will not open under the most severe normal operating conditions that could possibly be encountered, a group of two or more motors or one or more motors and other load(s), where each motor is provided with individual overload protection and the overcurrent device is sized according to NEC 430.52 to protect the smallest motor of the group, one branch circuit is all that's required.

NEC 430.53(C) - Other Group Installations - As with NEC 430.53(A) and (B) this section also allows two or more motors of any size or a combination of motors and other loads to be connected to one branch circuit. Motor controllers, overcurrent and overload protection devices may be installed as a listed (accepted) factory assembly or field-installed as separate assemblies.

As summarized, each motor overload device [or motor controller] must be listed for group installation with a specified maximum rating of fuse, inverse time circuit breaker [circuit breaker] *or* both *or* selected such that the ampere rating of the overcurrent protective device does not exceed that permitted by NEC 430.52 for that individual motor overload [controller] and corresponding motor load. Each circuit breaker is listed and is of the inverse time type. Fuses and inverse time circuit breakers protecting the branch circuit must have a rating not exceeding the specification of Table 430.52 for the highest rated motor plus the sum of other motors and the rating of other loads. The provisions of NEC 240.4(B) can be used if needed. Where overload relays protect the smallest motor of the broup, branch-circuit fuses or inverse time circuit breakers must comply with NEC 430.40. For other than nonmotor loads, overcurrent protection shall be in accordance with Parts I through VII of Article 240.

NEC 430.53(D) - Single Motor Taps - As described in NEC 430.53(A) - (C), for group installations, the conductors of any tap supplying a single motor shall not be required to have individual branch-circuit overcurrent protection devices, based on the following conditions:

(1) Conductors to the motor must have an ampacity not less than the branch-circuit conductors.

(2) Conductors that supply the motor must have an ampacity no less than one-third (1/3) of the branch-circuit conductors and comply with NEC 430.22, which require the ampacity of branch-circuit conductors to be 125 percent of the full-load current of the motor. The conductors to the motor overload device must not exceed 25 feet in length and be protected from physical damage by being enclosed in an approved raceway or by use of other approved means.

(3) Conductors from the branch-circuit overcurrent protection device to a listed manual motor controller additionally marked "Suitable for Tap Conductor Protection in Group Installations", are allowed to have an ampacity no less than one-tenth (1/10) of the rating of the branch-circuit overcurrent protection device.

Conductors from controller to motor must have an ampacity in accordance with NEC 430.22. Conductors from the branch-circuit overcurrent protection device to the controller must be suitably protected from physical damage, enclosed by an enclosed controller or raceway extending no more than 10 ft or have an ampacity not less than the branch-circuit conductors.

APPLYING THE NEC

Table 430.52 is used to establish the maximum rating or setting of a motor's branch-circuit overcurrent device. NEC 430.6(A)(1) states that the full-load current values provided in Tables 430.247 - 430.250, must be used for determining the ampere rating of a motor's branch-circuit overcurrent device, instead of the actual current rating marked on the motor's nameplate. Concurring with NEC 430.52(C)(1), the contents of Table 430.52 provides a listing of percent multipliers for increasing a motor's full-load current. The first column of the table identifies the various types of motors that are subject to the provisions of Table 430.52. The remaining four columns list a combination of overcurrent devices consisting of fuses and circuit breakers. Each set of percent multipliers per column represents the value a motor's full-load current is increased with regards to determining the rating or setting of the required overcurrent device. When the calculated values for branch-circuit overcurrent devices, as derived from Table 430.52, does not correspond to a standard size overcurrent device per NEC 240.6(A) the next higher standard size overcurrent device is allowed. Where the selected size of a protection device is insufficient for the starting current of a motor, NEC 430.52(C)(1), *Exception* No. *2* and 430.52(C)(3), *Exception* No. *1* permits the use of a larger size overcurrent device.

When sizing the branch-circuit overcurrent device for single motors the motor's full-load current must be known. Reference the applicable percent multiplier of Table 430.52 per motor type and desired overcurrent device. Increase the motor's full-load current by the applicable percent multiplier to size the motor's overcurrent device. Refer to the following procedures,

1. Determine the motor's full-load current as appropriately listed per Tables 430.247 - 430.250.

2. Multiply the selected full-load current by the applicable percent multiplier per Table 430.52 to obtain the rating or setting of the motor's branch-circuit overcurrent device. The next higher overcurrent device permitted if needed.

3. If the overcurrent device is insufficiently sized to allow the starting current of the motor, apply NEC 430.52(C)(1), *Exception No. 2* or 430.52(C)(3), *Exception No. 1*.

430.52(C)(1) - In Accordance with Table 430.52 (Rating or Setting) - *Exception Nos. 1* and *2*
Table 430.52 - Maximum Rating or Setting of Motor Branch-Circuit Short Circuit and Ground-Fault Protective Devices (37. - 38.) [2]

37. A 120V, 1½HP DC motor is required to be protected by an instantaneous trip circuit breaker. Determine the setting of the breaker.

Application

Step 1: Determine the motor's full-load current as appropriately listed per Tables 430.247 - 430.250.

Table 430.247 lists the full-load current values for DC (direct-current) motors. At 120V a 1½HP motor list a full-load current of 13.2A.

Step 2: Multiply the selected full-load current by the applicable percent multiplier per Table 430.52 to obtain the rating or setting of the motor's branch-circuit overcurrent device. The next higher overcurrent device permitted if needed.

As referenced in Table 430.52, a percent multiplier of 250 percent (2.50) is required to increase the DC motor's full-load current to determine the rating of the instantaneous trip circuit breaker. As a result,

$$13.2A \times 2.50 = 33A$$

NEC 430.52(C)(1), *Exception* No. *1* allows the use of the next higher size circuit breaker when an overcurrent device's rating does not correspond to a standard size. Referring to NEC 240.6(A) which lists the standard ampere ratings for overcurrent devices, a 35A instantaneous trip circuit breaker is the next standard size listed.

Step 3: If overcurrent device insufficiently sized to allow starting current of motor, apply NEC 430.52(C)(1), *Exception No. 2* or 430.52(C)(3), *Exception No. 1*.

In the event the instantaneous trip circuit breaker experience nuisance tripping due to being insufficiently sized to allow the starting current of the motor, NEC 430.52(C)(3), *Exception No. 1* permits the setting of such circuit breaker to be increased up to 1300 percent of the motor's full-load current for *other than* Design B energy-efficient motors which in this case applies to the DC motor. Although allowed, the intent was never to instantaneously go from sizing the motor's overcurrent device from 250 percent per Table 430.52 to 1300 percent, the intents were meant to be a gradual process.

In farther reading of the *Exception* it also states that the trip settings of an instantaneous trip circuit breaker above 800 percent for *other than* a Design B energy-efficient motor shall be permitted where demonstrated by an engineering evaluation. Therefore in limiting the setting of the instantaneous trip device, at 800 (8) percent the sizing of the circuit breaker would amount to,

$$13.2A \times 8 = 105.6A$$

which limits the setting of the instantaneous trip circuit breaker to 100A. If the setting of the circuit breaker was increased up to 1300 percent (13) it's maximum setting would be limited to 150A (13.2A x 13 = 171.6A).

38. Observe Figure 430.52(C)(1)-38. Determine the minimum size overcurrent devices needed for each motor's protection per given enclosure. Although the motors will be protected by either dual element time-delay (DETD) fuses or inverse time circuit breakers (ITCB) consider both devices for each motor. This question is a continuation of question No. 34.

240/120V - 3Ø - 4W Feeder

FEEDER CONDUCTORS - Use 3-2/0 AWG THWN Cu conductors per line OR 2-2/0 AWG (L₁ & L₃) and 1-2 AWG (L₂) THWN Cu conductors.

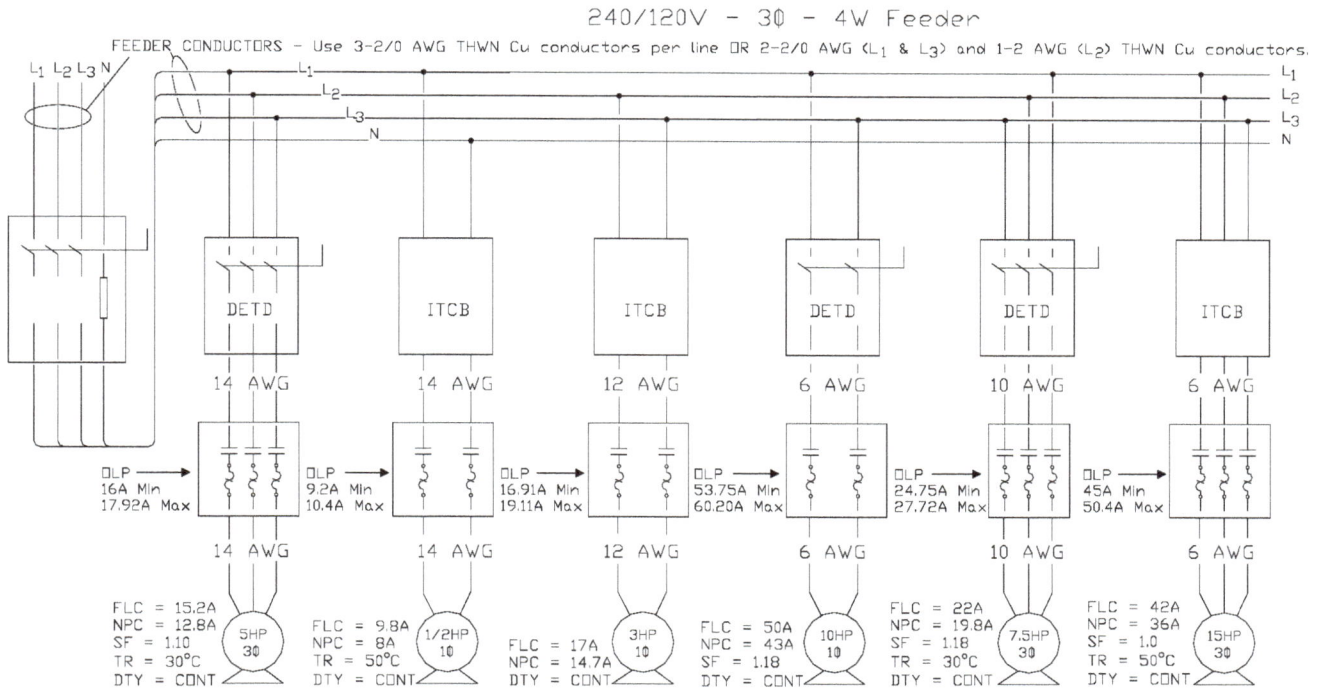

Figure 430.52(C)(1)-38

Application

Step 1: Determine the motor's full-load current as appropriately listed per Tables 430.247 – 430.250.

Step 2: Multiply the selected full-load current by the applicable percent multiplier per Table 430.52 to obtain the rating or setting of the motor's branch-circuit overcurrent device. The next higher overcurrent device permitted if needed.

Motor	FLC	PM	Dual Element Time-Delay Fuse	PM	Inverse Time Circuit Breaker
5HP - 3ϕ	15.2A	x 1.75 =	26.60A	x 2.50 =	38.00A
½HP - 1ϕ	9.8A	x 1.75 =	17.15A	x 2.50 =	24.50A
3HP - 1ϕ	17.0A	x 1.75 =	29.75A	x 2.50 =	42.50A
10HP - 1ϕ	50.0A	x 1.75 =	87.50A	x 2.50 =	125.00A
7.5HP - 3ϕ	22.0A	x 1.75 =	38.50A	x 2.50 =	55.00A
15HP - 3ϕ	42.0A	x 1.75 =	73.50A	x 2.50 =	105.00A

The minimum size overcurrent devices per motor are as follows. Remember NEC 430.52(C)(1), *Exception No. 1* allows the use of the next higher size overcurrent device when an overcurrent device's rating does not correspond to a standard size. Refer to NEC 240.6(A) for standard rating of overcurrent devices.

Motor	Minimum Dual Element Time-Delay Fuse	Minimum Inverse Time Circuit Breaker
5HP - 3φ	30A	40A
½HP - 1φ	20A	25A
3HP - 1φ	30A	45A
10HP - 1φ	90A	125A
7.5HP - 3φ	40A	60A
15HP - 3φ	80A	110A

In question No. 18., each motor's branch-circuit conductors were determined. Comparing the branch-circuit conductors to the ratings of the overcurrent devices, would make one assume there's a mistake because the overcurrent devices are sized much higher than the ampacity of the branch-circuit conductors. For motor installations this is common practice, the thing to keep in mind is, the branch-circuit conductors are provided with dual means of protection.

The overload protection determined in question Nos. 32. or 35. will allow current to flow to the motor as long as the current does not exceed the calculated values for a continuous period of time. The calculated values for each motor's overload protection relative to the ampacity of the branch-circuit conductors, clearly explains this means of protection. The overcurrent protection devices determined in this question, whether dual element-time-delay fuses or an inverse time circuit breaker, will protect the motor's branch-circuit conductors against short-circuit and ground-faults, which is actually what the contents of NEC 430.51 states.

Step 3: If overcurrent device insufficiently sized to allow starting current of motor, apply NEC 430.52(C)(1), *Exception No. 2* or 430.52(C)(3), *Exception No. 1*.

According to NEC 430.52(C)(1), *Exception No. 2(b)* permits the rating of a time-delay (dual element) fuse to be increased but in no case exceed 225 percent (2.25) of the full-load current of a motor while *Exception No. 2(c)* permits the rating of an inverse time circuit breaker to be increased but in no case exceed 400 percent (4) of the full-load current of a motor of 100 amperes or less or no more than 300 percent (3) of the full-load current of a motor greater than 100 amperes.

As it pertains to *Exception No. 2(c)* because none of the motor full-load currents are not rated higher than 100 amperes the inverse time circuit breaker's increase cannot exceed 400 percent. Applying the same approach as in **Step 2**.

Motor	FLC	PM	Dual Element Time-Delay Fuse	PM	Inverse Time Circuit Breaker
5HP - 3φ	15.2A	x 2.25 =	34.20A	x 4 =	60.80A
½HP - 1φ	9.8A	x 2.25 =	22.05A	x 4 =	39.20A
3HP - 1φ	17.0A	x 2.25 =	38.25A	x 4 =	68.00A

$$10HP - 1\phi \quad 50.0A \quad x \quad 2.25 = 112.50A \qquad x \quad 4 = 200.00A$$
$$7.5HP - 3\phi \quad 22.0A \quad x \quad 2.25 = 49.50A \qquad x \quad 4 = 88.00A$$
$$15HP - 3\phi \quad 42.0A \quad x \quad 2.25 = 94.50A \qquad x \quad 4 = 168.00A$$

The maximum size overcurrent devices per motor are as listed below. Although NEC 430.52(C)(1), *Exception No. 1* allows the use of the next higher size overcurrent device when an overcurrent device's rating does not correspond to a standard size, *Exception Nos. 2(b)* and *2(c)* does not. Refer to NEC 240.6(A) for standard rating of overcurrent devices.

Motor	Maximum Dual Element Time-Delay Fuse	Maximum Inverse Time Circuit Breaker
5HP - 3ϕ	30A	60A
½HP - 1ϕ	20A	35A
3HP - 1ϕ	35A	60A
10HP - 1ϕ	110A	200A
7.5HP - 3ϕ	45A	80A
15HP - 3ϕ	90A	150A

Take notice of how the maximum dual element-time delay fuse ratings are the same as the minimum fuse ratings for the 5HP-3ϕ and ½HP-1ϕ motors.

430.52(C)(1) - In Accordance with Table 430.52 (Rating or Setting)
Table 430.52 - Maximum Rating or Setting of Motor Branch-Circuit Short Circuit and Ground-Fault Protective Devices - **430.52(C)(1)**, *Exception Nos. 1* and *2*
430.52(C)(4) - Multispeed Motor (Rating or Setting)

39. Refer to questions No. 17., 20. and 21. If the motors given in each question requires being protected by nontime-delay fuses, what size fuses are needed for each motor?

Application

Step 1: Determine the motor's full-load current as appropriately listed per Tables 430.247 - 430.250.

Question No. 17. - Three-phase, 230V, 50HP synchronous motor
FLC = 114.4A

Three-phase, 460V, 100HP synchronous motor
FLC = 126.25A

Question No. 20. - Three-phase, 230V, 75HP multi-speed motor
NPC (smallest winding) = 59A

According to NEC 430.52(C)(4) the nameplate current of the smallest winding of a multi-speed motor must be used when the *Exception* to 430.52(C)(4) is not applied.

Question No. 21. - Three-phase, 460V, 125HP part-winding motor
FLC = 156A

Three-phase, 460V, 125HP wye-start, delta-run motor
FLC = 180A

Step 2: Multiply the selected full-load current by the applicable percent multiplier per Table 430.52 to obtain the rating or setting of the motor's branch-circuit overcurrent device. The next higher overcurrent device permitted if needed.

The percent multipliers, for the five given motors, according to Table 430.52 requires the same 300 percent (3.00) value, therefore,

114.4A x 3.00 = 343.2A
126.25A x 3.00 = 378.75A

59A x 3.00 = 177A

156A x 3.00 = 468A
180A x 3.00 = 540A

NEC 430.52(C)(1), *Exception No. 1* allows the use of the next higher size fuse when an overcurrent device's rating does not correspond to a standard size. Again, referring to NEC 240.6(A) which list the standard ampere ratings for overcurrent devices, the following nontime-delay fuses are required,

3ϕ, 230V, 50HP synchronous motor, at 343.2A, use 3-350A fuses
3ϕ, 230V, 100HP synchronous motor, at 378.75A, use 3-400A fuses
3ϕ, 230V, 75HP multi-speed motor, at 177A, use 3-175A fuses*

*If the next higher size fuse was used, it would be in violation of NEC 430.52(C)(4) because it would exceed the applicable percentage of the nameplate current rating of the smallest winding protected. If a 200A fuse was used it would exceed the 300 percent multiplier by approximately 39 percent (200A/59A = 3.3898 [x 100] = 338.98 percent – 300 percent = 38.98 percent).

Just as explained and resolved in question No. 2., the same approach must be taken here in determining the size nontime-delay fuses needed for the part-winding motor. Again, referring to NEC 430.4 which requires each motor-winding connection to have branch-circuit short-circuit and ground-fault (overcurrent) protection rated at *not more than one-half* of that *specified by* NEC 430.52.

At 468A, 500A fuses would be required. To determine the branch-circuit short-circuit and ground-fault protection, the rating of the permitted device must be multiplied by one-half (½).

Therefore,

$$500A \times \tfrac{1}{2} \ or \ 500A \ / \ 2 = 250A$$

Based on the calculation, to protect the motor, 250A fuses are required to protect each motor-winding connection.

3ϕ, 460V, 125HP wye-start, delta-run motor, at 540A, use 3-600A fuses

Step 3: If overcurrent device insufficiently sized to allow starting current of motor, apply NEC 430.52(C)(1), *Exception No. 2* and/or 430.52(C)(3), *Exception No. 1.*

In the event the above nontime-delay fuses prove to be insufficient, NEC 430.52(C)(1), *Exception No. 2(a)* allows the rating of the fuse to be increased up to but not exceeding 400 percent of the motor's full-load current providing the fuse does not exceed 600A. Take a look,

$$114.4A \times 4.00 = \textbf{457.6A}$$

The three-phase, 230V, 50HP synchronous motor can only be protected at most by 450A fuses. Any fuse rated above **457.6A** would exceed the 400 percent limit.

$$125.25A \times 4.00 = \textbf{501A}$$

The three-phase, 460V, 100HP synchronous motor can only be protected at most by 500A fuses. Any fuse rated above **501A** would exceed the 400 percent limit.

The three-phase, 230V, 75HP multi-speed motor *cannot* be protected beyond the 175A fuses based upon the use of the smallest winding. Again, refer to **Step 2** (*).

$$156A \times \tfrac{1}{2} \times 4.00 = \textbf{312A}$$

Based upon the 400 percent limit, the three-phase, 460V, 125HP part-winding motor can only be protected per winding connection by 300A fuses.

$$180A \times 4.00 = \textbf{720A}$$

The three-phase, 460V, 150HP wye-start, delta-run motor is already being protected by the maximum allowable set of fuses. The rating of the fuses can never exceed 600A per NEC 430.52(C)(1), *Exception No. 2(a)*.

430.52(D) - Torque Motors (Rating or Setting for Individual Motor Circuit)

40. What size overcurrent protection device is required to protect the torque motor in question No. 5.?

As it pertains to torque motors, NEC 430.6(B) concludes by stating, "the ampere rating of motor branch- circuit short-circuit and ground-fault protection in accordance with NEC 430.52(B)". NEC 430.52(B) requires such protection to be capable of carrying the starting current of the motor.

Now, according to NEC 430.52(D) torque motor branch circuits shall be protected at the motor nameplate current rating in accordance with NEC 240.4(B). The provisions of NEC 240.4(B) pertain to overcurrent devices rated 800 amperes or less. Based on the 928A nameplate current of the torque motor, as a minimum, a 1000A overcurrent device is required to protect the motor.

Unlike NEC 240.4(B), the provisions of NEC 240.4(C) pertain to overcurrent devices rated over 800 amperes which covers the required 1000A overcurrent device for the torque motor.

In this case, because the torque motor's minimum overcurrent device is rated over 800 amperes the provisions of NEC 240.4(C) must be applied. In question No. 5. the calculated ampacity of the required branch-circuit conductors for the torque motor amounted to 1160 amperes. Per NEC 240.4(C), the ampacity of the conductors being protected by an overcurrent device rated over 800 amperes are required to be equal to or greater than the rating of the overcurrent device.

Therefore, the torque motor's branch-circuit conductors when selected must be either equal to or exceed the 1160A calculated ampacity which will in terms surpass the rating of the 1000A overcurrent device. In the event the overcurrent device proves to be incapable of carrying the starting current of the motor, NEC 430.52(B) must be adhered to.

FEEDER-CIRCUIT OVERCURRENT PROTECTION

ACCORDING TO THE NEC

The following NEC sections and tables are applicable for understanding how to size motor feeder circuit short-circuit and ground-fault protection devices per Part IV of Article 430.

FOR MOTOR LOADS ONLY

NEC 430.62(A) - Specific Load (Rating or Setting - Motor Load) - A feeder conductor supplying a fixed motor load(s) where the size of the feeder conductor is calculated according to NEC 430.24 shall be protected by an overcurrent protection device based on the largest overcurrent protection device provided for any motor supplied by the feeder conductor [with regards to the maximum permitted value designated for a specific overcurrent protection device according to NEC 430.52 or 440.22(A) for hermetic refrigerant motor-compressors] plus the sum of the full-load currents of the other motors of the group. When two or more branch circuits supplied by the feeder conductor have the same overcurrent protection device rating, one of the protection devices shall be considered the largest when calculating a feeder conductor's overcurrent protection device.

NEC 430.62(B) - Other Installations - Feeder conductors having an ampacity greater than required by NEC 430.24 are allowed to have the feeder overcurrent protection device sized according to the ampacity of the feeder conductors. The intent of this allowance is to accommodate future load additions or changes to existing feeder conductors.

APPLYING THE NEC

NEC 430.62(A) defines the means for determining the rating or setting of an overcurrent protection device for a common feeder conductor supplying a group of motors. According to this reference, the feeder conductor's overcurrent device is based upon the largest standard size overcurrent device utilized by any motor or motor-compressor within the group of motors and the actual full-load current rating of all remaining motors supplied by the feeder conductors. In the event two or more motors share the same size overcurrent device, one of the overcurrent devices must be considered the largest for calculating the feeder conductor's overcurrent protection. Unlike NEC 430.52(C)(1), *Exception No. 1*, the next higher standard size overcurrent device is not allowed when the calculated value for a feeder-circuit's overcurrent protection does not correspond to a standard size overcurrent device.

When sizing the overcurrent device for feeder conductors supplying a group of motor loads, the largest rated overcurrent device of the group is combined with the full-load current ratings of all remaining motors to derive the rating or setting of the feeder conductor's overcurrent device. Refer to the following procedures:

1. Determine the largest rated overcurrent device (whether fuse or circuit breaker) per motor among the group of motors.

2. Add the largest rated overcurrent device and the full-load current ratings of all remaining motors.

3. Use a standard rated overcurrent device not exceeding the derived total.

430.62(A) - Specific Load (Rating or Setting - Motor Load)/430.28 - Feeder Taps (41. - 43.) [3]

41. Using the data in Figure 430.62(A)-41, determine the size of the feeder conductor's overcurrent device if dual-element time delay fuses were used. This question is a continuation of question No. 38.

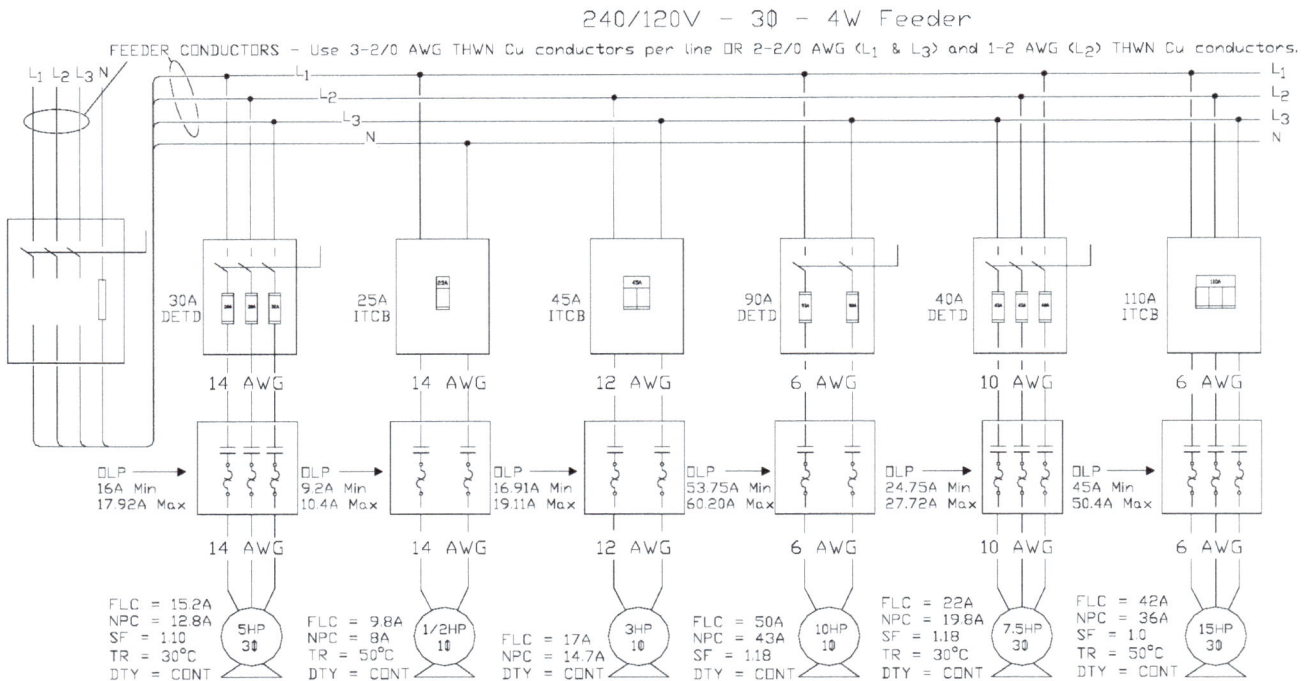

240/120V - 3Ø - 4W Feeder

Figure 430.62(A)-41

Application

Step 1: Determine the largest rated overcurrent device (whether fuse or circuit breaker) per motor among the group of motors.

Remember back in question No. 25. how the size of the feeder conductors were determined based on the application of "all motors" followed by the "per line" alternative. In this question the fundamentals of both methods are considered. Observe,

LARGEST OVERCURRENT DEVICE (all motors)

110A Inverse Time Circuit Breaker

LARGEST OVERCURRENT DEVICE (per line)

L_1, L_2 and $L_3 =$ **110A** Inverse Time Circuit Breaker

Step 2: Add the largest rated overcurrent device and the full-load current ratings of all remaining motors.

"all motors"

$$110A + 15.2A + 9.8A + 17.0A + 50.0A + 22.0A = \mathbf{224A}$$

"per line"

$$L_1 = \mathbf{110A} + 15.2A + 9.8A + 50.0A + 22.0A = \mathbf{207A}$$
$$L_2 = \mathbf{110A} + 15.2A + 17.0A + 22.0A + = \mathbf{164.2A}$$
$$L_3 = \mathbf{110A} + 15.2A + 17.0A + 50.0A + 22.0A = \mathbf{214.2A}$$

Step 3: Use a standard rated overcurrent device not exceeding the derived total.

Regardless of which method chosen ("all motors" or "per line") NEC 430.62(A) does not permit the use of the next higher size overcurrent device.

As a result, the "all motors" method in this situation (and other likewise situations) will always prove, for the most part, to be the easiest and the most practical approach for sizing feeder conductors and their overcurrent device.

Because the primary source of overcurrent protection for the feeder per Figure 430.62(A)-41 illustrates the use of a fusible disconnect switch which serves as the disconnecting means and the enclosure for the overcurrent device protecting the feeder conductors., the "all motors" method would require the use of 200A dual element time fuses (DETD) as a whole while the "per line" method would require 200A DETD fuses to be utilized for lines L_1 and L_3 and the use of a 150A DETD fuse for line L_2.

42. Referring to Figure 430.62(A)-42, determine the size overcurrent protection required for the feeder conductor.

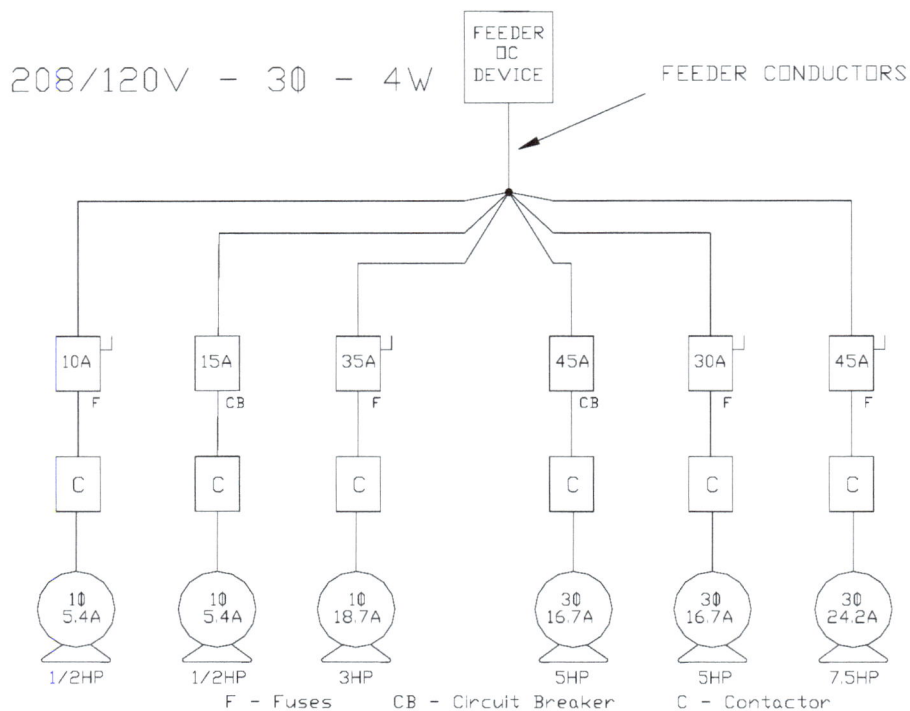

Figure 430.62(A)-42

Application

Step 1: Determine the largest rated overcurrent device (whether fuse or circuit breaker) per motor among the group of motors.

Of the six given motors, two of the motors are protected by overcurrent devices having the same rating. According to NEC 430.62(A) when two or more motors supplied by the same feeder conductors are protected by overcurrent devices of the same rating, one of the devices must be considered the largest for calculating the feeder conductor's overcurrent device.

Consider the two motors having the largest overcurrent devices. Although the difference in horsepower rating is irrelevant, both motors are protected by different type overcurrent devices yet both devices are rated for 45A.

Step 2: Add the largest rated overcurrent device and the full-load current ratings of all remaining motors.

To make this situation interesting, let's calculate the feeder conductor's overcurrent device by considering both overcurrent devices individually.

Condition A - 5HP motor protected by 45A device (circuit breaker)

$$45A + 5.4A + 5.4A + 18.7A + 16.7A + 24.2A = 115.4A$$

Condition B - 7.5HP motor protected by 45A device (fuses)

$$45A + 5.4A + 5.4A + 18.7A + 16.7A + 16.7A = 107.9A$$

Step 3: Use a standard rated overcurrent device not exceeding the derived total.

Based on the results of Condition A, a 110A overcurrent device is required, whereas with Condition B, a 100A overcurrent device is required. The question now is, "which one do I use?", and the answer is "either", NEC 430.62(A) simply leaves that decision up to the user.

43. This question reference the application of NEC 430.28. Observing Figure 430.28-43A, what size tap conductors are required to supply each individual motor? Assume all tap conductors will be enclosed in raceway. Use the same type insulation as the feeder conductors. The temperature rating of all terminating means is 75°C. This question is a continuation of question No. 41.

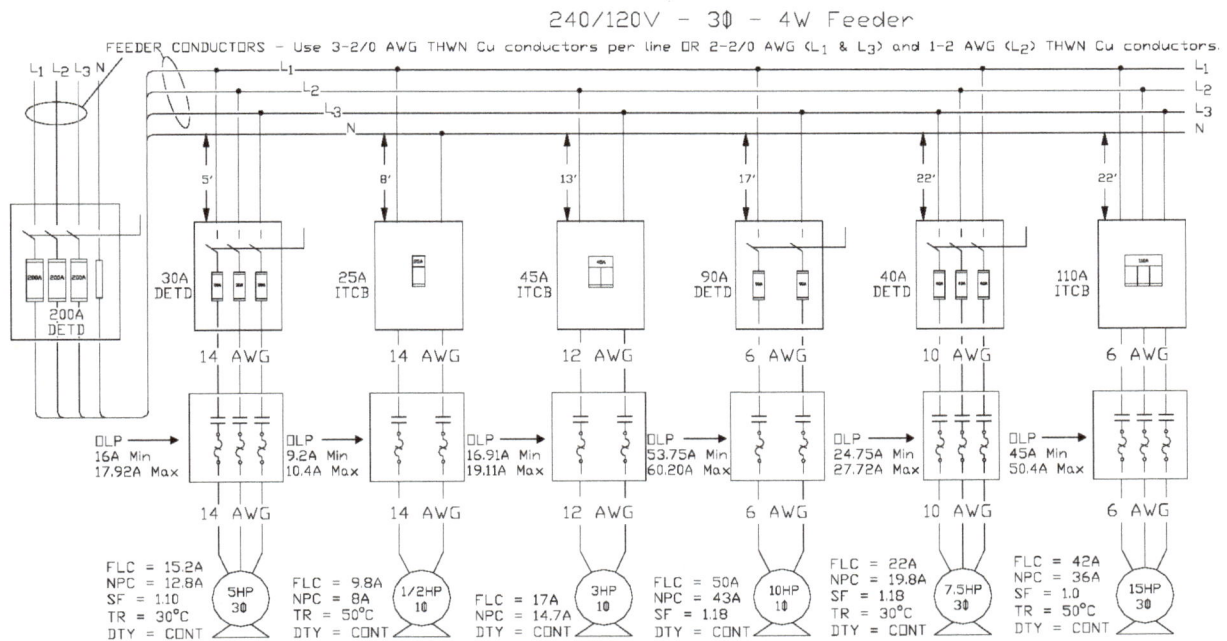

Figure 430.28-43A

When two conductors having the same ampacity and of the same size are joined together the connection is called a splice providing the same overcurrent protection for spliced conductors offers identical protection for both conductors.

In contrast and in most situations, when a conductor of a smaller size (lower ampacity) is joined to a conductor of a larger size (higher ampacity) the connection is called a tap. Where applicable, the smaller size conductor is called a *tap conductor* providing the overcurrent protection ahead of the larger conductor does not provide equal protection for the smaller conductor. Overcurrent protection for a tap conductor is not required at the point of a tap connection when certain conditions of the NEC are met. Per NEC 240.2 a *tap conductor* is defined as a conductor, other than a service conductor that has overcurrent protection ahead of its supply (in this case the 200A feeder overcurrent protection) that exceeds the value permitted for similar conductors that are protected as described elsewhere in NEC 240.4.

NEC 240.21(B)(1) - (5) provide general conditions for feeder taps where nonmotor loads are involved. NEC 430.28 provides specific conditions for feeder taps involving motor loads when tap conductors will extend up to and surpass 10' but never beyond 25' without exception.

Referring to Figure 430.28-43A, we see that the length from the point of tap for the six motors range from 5 to 22 feet, therefore the tap conductors for each motor must be sized according to the provisions of NEC 430.28.

Tap conductors up to 10'

Tap conductors up to 10' must be enclosed by means of a controller or raceway and the rating of the feeder conductor's overcurrent device preceding the tap conductor cannot be greater than 1000 percent (10.0 times) of the rated ampacity of the tap conductor.

The ½ and 5 horsepower motors have tap conductors under 10' that are assumed to be enclosed in raceway. To determine the size tap conductors required for both motors let's first consider each motor's branch-circuit conductors. The 14 AWG THWN copper conductors @ 75°C that supply both motors have an ampacity of 20 amps per Table 310.15(B)(16). In order to size the tap conductors, the rating of the overcurrent device (200A) protecting the feeder conductors cannot be more than 1000 percent (10 times) of the ampacity of the tap conductors where when calculated (20A x 10 = 200A) the product reflects the 1000 percent limitation not being exceeded.

Because the provisions of NEC 430.28(1) are met, the same size conductors (14 AWG) can also be used as tap conductors for the ½HP and 5HP motors.

Tap conductors beyond 10' up to 25'

Tap conductors ranging within these lengths must also be enclosed in raceway or be suitably protected from physical damage. The ampacity of the tap conductors must be at least equal to one-third (1/3) of the ampacity of the feeder conductors (58.3A).

All remaining motors being supplied by the 2/0 AWG THWN copper conductors that extend beyond 10' up to 25' must have an ampacity that is either equal to or exceeds 58.3A (1/3 of the feeder conductor's ampacity - 175A x 1/3 = 58.3A). As a result, 6 AWG THWN copper (65A) conductors can also serves as the tap conductors for the 10 and 15 horsepower motors and must, as a minimum, also serve as the tap conductors for the 3 and 7.5 horsepower motors.

Even if applying the "alternate" method where a 1 AWG THWN copper conductor (130A) is used as the feeder conductor L_2, the ampacity of the 6 AWG tap conductors still exceeds one-third (130A/3 = 43.3A) of the ampacity of the 1 AWG conductors.

Figure 430.28-43B reflects the illustrated conclusion of questions 18., 25., 32., 35., 38., 41. and 43A.

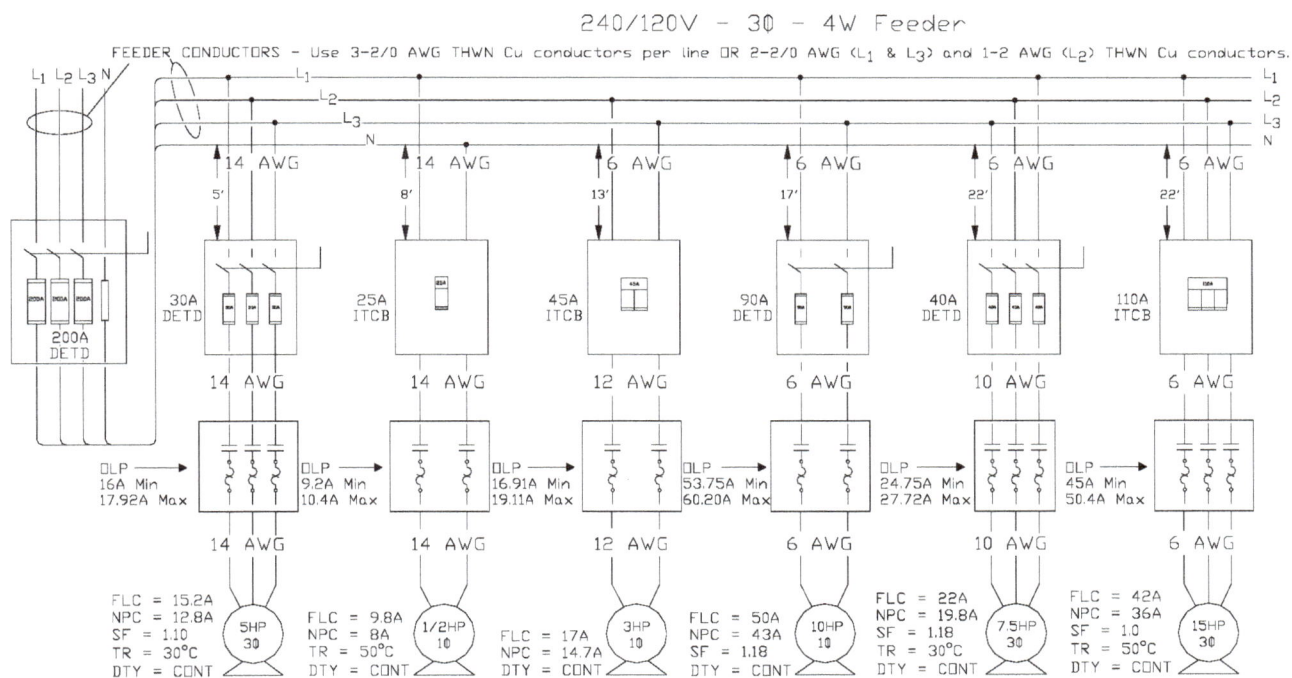

Figure 430.28-43B

Now as far as the sizing of the system's neutral conductor is concerned based on the only neutral load of all motors, the system's neutral conductor could be sized based on the full-load current of the ½HP motor at 9.8A. However, as earlier mentioned NEC 215.2(A)(2) and 250.122 requires it to be sized no smaller than the provisions of NEC and Table 250.122. Therefore, based on the 200A overcurrent device protecting the feeder conductors, per Table 250.122, the neutral conductor can be sized no smaller than a 6 AWG copper or a 4 AWG aluminum as a minimum or the neutral conductor can be sized the same as the largest feeder conductor.

ACCORDING TO THE NEC

FOR MOTOR(S) and NONMOTOR LOADS

NEC 430.63 - Rating or Setting - Motor Load and Other Load(s) - Feeder conductors supplying a motor load along with other load(s) must have a feeder overcurrent device rating not less than that required for the sum of the other load(s) *plus* for a single motor, the rating permitted in NEC 430.52 *or* for a single hermetic refrigerant motor-compressor, the rating permitted in NEC 440.22 *or* for a combination of two or more motors, the rating permitted in NEC 430.62.

APPLYING THE NEC

NEC 430.63 defines the means for determining the rating or setting of an overcurrent protection device for a feeder conductor supplying other (nonmotor and remaining motor) loads. Where this occurs, the sum of the other loads and the *largest* single (branch-circuit short-circuit ground-fault protection) overcurrent device whether it originates from **(1)** a (standard) single motor as permitted per NEC 430.52 *or* **(2)** from a single hermetic refrigerant motor-compressor as permitted per NEC 440.22 *or* **(3)** from the *largest* single (standard or hermetic refrigerant motor-compressor) motor (where two or more) and the full-load currents of all remaining motors as permitted per NEC 430.62. Regardless of requirement, the resulting feeder overcurrent device must have a rating that is either equal to or exceed the sum of the other loads and the largest selected single overcurrent device.

When sizing the feeder overcurrent device, affiliate motor's full-load current(s) must be known. Reference the applicable percent multiplier of Table 430.52 per motor type and the desired overcurrent device. Increase the motor's full-load current by the applicable percent multiplier to size motor's overcurrent device.

Applications

1. Determine the other load(s) (nonmotor-continuous and noncontinuous).

2. Determine the largest rated overcurrent device (whether fuse or circuit breaker) per single motor or required among a group of motors (two or more) per NEC 430.62.

3. Total other load(s) and the largest selected overcurrent device (steps **1.** and **2.**).

4. Use an overcurrent device rated not less than the derived total.

430.63 - Rating or Setting - Motor Load and Other Load(s) (44. - 46.) [3]

44. Refer to question No. 26. Assume the 25HP motor is protected by 50A time-delay fuses, what size overcurrent device is needed to protect the feeder conductor?

Applications

Step 1: Determine the other load(s) (nonmotor-continuous and noncontinuous).

The other (transformers) load of question No. 26. totals 90.2A.

Step 2: Determine the largest rated overcurrent device (whether fuse or circuit breaker) per single motor or required among a group of motors (two or more) per NEC 430.62.

The only motor load in question No. 26. is the 25HP (standard) motor. As assumed it is protected by 50A time-delay fuses.

Step 3: Total other load(s) and the largest selected overcurrent device (steps **1.** and **2.**).

90.2A (other load) + 50A (largest rated overcurrent device) = 140.2A

Step 4: Use an overcurrent device rated not less than the derived total.

Based on the derived total a 150A overcurrent device is required, be it fuses or circuit breaker to protect the feeder conductors.

45. A 400A three-phase sub-panelboard rated for 208/120V is being supplied by a feeder from a 1200A main distribution panelboard. The sub-panelboard supplies a string of 400W metal halide light fixtures (208V) totaling 60A, 88 duplex receptacles (120V), two three-phase 15kW electric furnaces (208V) and a 10HP three-phase air handler motor. The motor is required to be protected by an inverse time circuit breaker (ITCB) at maximum rating. Based on the given information, what size inverse time circuit breaker is required to protect the feeder?

Application

Step 1: Determine the other load(s) (nonmotor-continuous and noncontinuous).

Lighting (continuous - assumed to operate more than 3 hours)

60A x 1.25 = 75A

Receptacles (noncontinuous) - 88 duplex receptacles

[NEC 220.14(I), 220.44, Table 220.44]
88 x 180VA = 15,840VA

$$
\begin{array}{lll}
\text{lst 10kVA} & = & 10,000\text{VA} \\
\text{(Remaining) } 5840\text{VA x .50} = & & \underline{2,920\text{VA}} \\
& & 12,920\text{VA}
\end{array}
$$

12,920VA / 208V x √3 = 35.86A (√3 = 1.732)

Heating Equipment - 2 units @15kW, 3φ

15,000W x 2 / 208V x √3 = 83.27A

Other Loads Total - 75A + 35.86A + 83.27A = 194.13A

Step 2: Determine the largest rated overcurrent device (whether fuse or circuit breaker) per single motor or required among a group of motors (two or more) per NEC 430.62.

To size the overcurrent device for the 10HP motor, NEC 430.52(C)(1), *Exception No. 2(c)* and Table 430.250 is referenced. Based on both references,

Percentage of Full-Load Current (maximum) = 400 [4] (ITB) FLC = 30.8A

$$30.8A \times 4 = 123.2A$$

Because the maximum rating of the inverse time circuit breaker cannot exceed 400 percent [NEC 430.52(C)(1), *Exception No. 2(c)*] of the motor's full-load current the rating of the breaker is limited to 110A. As a result, a 110A inverse time circuit breaker is required to protect the 10HP motor.

Step 3: Total other load(s) and the largest selected overcurrent device (steps **1.** and **2.**).

$$194.13A \text{ (other loads)} + 110A \text{ (motor load)} = 304.13A$$

Step 4: Use an overcurrent device rated not less than the derived total.

Based on the derived total, a 350A inverse time circuit breaker is required to protect the feeder.

46. What size time delay fuses are needed to protect the feeder conductors in question No. 27. The motors are required to be protected by inverse time circuit breakers?

Application

Step 1: Determine the other load(s) (nonmotor-continuous and noncontinuous).

In question No. 27. the other loads are as follows,

Lighting	=	73.00A
Receptacles	=	48.96A
Kitchen Equipment	=	132.79A
Total other loads	=	254.75A

Step 2: Determine the largest rated overcurrent device (whether fuse or circuit breaker) per single motor or required among a group of motors (two or more) per NEC 430.62.

Referring back to NEC 430.63, subsection (3) describes the situation in question No. 27. being that there are more than two motors in this installation. Having considered the other loads, the required overcurrent device among the motors must originate from the largest single motor per highest full-load current and the full-load currents of all remaining motors as permitted per NEC 430.62.

In question No. 27. the single-phase 10HP motor was considered the largest rated motor based on its full-load current per NEC 430.17. Since all motors will be protected by the same type overcurrent device, the 10HP motor will also have the largest rated overcurrent device. Using Table 430.52 and the motor's full-load current the rating of the motor's overcurrent device is determined.

50A x 2.50 = 125A (10HP motor overcurrent device)

Applying the largest rated overcurrent device along with the full-load current ratings of other motors as listed in question No. 27. totals 252A.

125A + 17A + 40A + 28A + 42A = 252A (device not to exceed total)

According to NEC 430.62(A) a 250A overcurrent device is required based on the above calculation.

Step 3: Total other load(s) and the largest selected overcurrent device (steps **1.** and **2.**).

254.75A (nonmotor loads) + 250A (motor loads) = 504.75A

Step 4: Use an overcurrent device rated not less than the derived total.

Based on the derived total (504.75A) 600A time delay fuses are required.

430.72(B) - Conductor Protection (Overcurrent Protection)-*Exception No. 2*

47. Size 14 AWG copper conductors are used as the secondary conductors of a 240-120V, 1φ, 2W motor control transformer. If the transformer and secondary conductors are enclosed in a motor controller, can the conductors be protected by the overcurrent device protecting the transformer's primary side?

According to NEC 430.72(B), *Exception No. 2* conductors that are supplied by the secondary side of a single voltage, 2-wire single-phase transformer are permitted to be protected by overcurrent protection provided on the primary side of a transformer if the rating of the primary overcurrent protection does not exceed the value determined by multiplying the appropriate maximum rating of the overcurrent device for the secondary conductor from Table 430.72(B) by the secondary-to-primary voltage ratio.

Per **Column B** of Table 430.72(B), 14 AWG conductors can be protected by a 100A protective device if the conductors are within an enclosure. Therefore, based on the given requirements,

100A x 120V/240V = 50A

The secondary conductors can be protected by the overcurrent device protecting the transformer's primary side providing the device does not exceed 50A.

430.72(B)(1) - Separate Overcurrent Protection (Conductor Protection)

48. If 12 AWG copper motor circuit conductors required separate overcurrent protection, what size protective device would be required?

NEC 430.72(B)(1) reference the use of **Column A** of Table 430.72(B) for such requirements. **Column A** of Table 430.72(B) reference the use of **Note 1** of the table's footnotes for determining the required overcurrent device. **Note 1** states that the value specified in NEC 310.15 must be applied in the selection of the overcurrent device. NEC 310.15(A)(1) reference the use of the tables listed in NEC 310.15(B). Therefore, in compliance with Tables 310.15(B)(16) and 310.15(B)(17) based on the provisions given in the headings of the tables, and taking into consideration NEC 110.14(C)(1)(a), at 60°C per Table 310.15(B)(16) a 20A device is required and per Table 310.15(B)(17) a 30A device is required based on the use of 12 AWG copper conductors.

430.72(B)(2) - Branch-Circuit Overcurrent Protective Device (Conductor Protection)
(49. - 50.)[2]

49. If 8 AWG copper conductors were used as the conductors for a motor control circuit and were installed in an enclosure, what size overcurrent device would be required to protect the conductors? If the conductors extended beyond the enclosure?

Per **Column B** of Table 430.72(B) which reference **Note 2** for such conductors larger than 10, **Note 2** requires the value specified in Table 310.15(B)(17) for 60°C to be increased by 400 percent (4).

Per Table 310.15(B)(17), at 60°C the ampacity of an 8 AWG copper conductor is 60A. Therefore, 60A x 4 = 240A.

This requires the use of a 225A overcurrent device because the calculated value cannot exceed 400 percent. The next standard size overcurrent device per NEC 240.6(A) at 250A would exceed the 400 percent limitation.

Per **Column C** of Table 430.72(B) which reference **Note 3** for conductors larger than 10 AWG that will extend beyond an enclosure, Note 3 requires the value specified in Table 310.15(B)(16) for 60°C to be increased by 300 percent (3). Per Table 310.15(B)(16) at 60°C, the ampacity of an 8 AWG copper conductor is 40A. Therefore, 40A x 3 = 120A.

This requires the use of a 110A overcurrent device because the calculated value cannot exceed 300 percent. The next standard size overcurrent device per NEC 240.6(A) at 125A would exceed the 300 percent limitation.

In summary, the ampacity of the conductors installed in an enclosure was allowed to be taken from Table 310.15(B)(17) based on "free air" ampacity. This was allowed because the conductors are installed in an open enclosure opposed to an enclosed raceway which would not allow the conductors to dissipate (spread out) heat as an open enclosure would.

50. If the 12 AWG copper conductors in question No. 48. were either enclosed in an enclosure or extended beyond an enclosure of a motor controller, what size protective device would be required?

NEC 430.72(B)(2) reference the use of **Column B** of Table 430.72(B) when such conductors will be enclosed in an enclosure. According to **Column B**, the overcurrent device must not be rated higher than 120A.

NEC 430.72(B)(2) reference the use of **Column C** of Table 430.72(B) when such conductors will extend beyond an enclosure. According to **Column C**, the overcurrent device must be rated for a maximum of 60A.

430.72(C)(4) - Primary Less Than 2 Amperes (Control Circuit Transformer)

51. A motor control transformer has a rated primary current of 1.7A. What size primary overcurrent device is required to protect the control transformer?

According to NEC 430.72(C)(4), for control circuit transformers with a rated primary current less than 2 amperes, an overcurrent device rated or set at not more than 500 percent (5) of the rated primary current shall be permitted in the primary circuit. As a result,

$$1.7A \times 5 = 8.5A$$

Because the rating cannot exceed 500 percent, the overcurrent device cannot exceed 8.5A. Per NEC 240.6(A) the minimum size circuit breaker listed is rated for 15A. Per NEC 240.6(A) only fuses are rated less than 15-amperes. Based on the calculated results, a fuse rated for 6A must be used to protect the transformer's primary.

430.110(A) - General (Ampere Rating and Interrupting Capacity)

Refer to NEC 440.12(A)(1) and question No. 1. of Article 440 for similar application.

430.110(B) - For Torque Motors (Ampere Rating and Interrupting Capacity)

52. Determine the required ampere rating of a disconnecting means if used for the torque motor in question No. 5.

According to NEC 430.110(B) the disconnecting means for a torque motor shall have an ampere rating of at least 115 percent (1.15) of the motor's nameplate current. Per question No. 5. the nameplate current of the motor is 928A where, 928A x 1.15 = 1067.2A.

As a minimum, the disconnecting means must be rated no less than 1067.2A.

430.110(C)(1) - Horsepower Rating (For Combination Loads)

Refer to NEC 440.12(B)(1) and question No. 3. of Article 440 for similar application.

430.110(C)(2) - Ampere Rating (For Combination Loads)

Refer to NEC 440.12(B)(2) and question No. 4. of Article 440 for similar application.

430.110(C)(3) - Small Motors (For Combination Loads)

53. Per manufacturer's specifications, the full-load current of a three-phase 208V, 35HP motor is 103A. Based on the specified full-load current, can the motor's locked-rotor current be determined?

According to NEC 430.110(C)(3), where small motors are not listed in any of the full-load current Tables of Article 430, a motor's locked-rotor shall be assumed to be six times (6) the full-load current of the unlisted motor. As a result the motor's assumed locked-rotor current is:

$$103A \times 6 = 618A$$

430.122(A) - Branch/Feeder Circuit Conductors (Conductors - Minimum Size and Ampacity)

The branch-circuit conductors for an adjustable speed drive system must have an ampacity rating not less than 125 percent (1.25) of the rated input current to the power conversion equipment. Where an adjustable speed drive system includes the use of a bypass device refer to NEC 430.122(B). The branch-circuit conductor requirements of NEC 430.122(A) apply the same procedure as used for continuous loads, motor loads, motor-compressor loads and others.

430.122(B) - Bypass Device (Conductors – Minimum Size and Ampacity

The provisions of NEC 430.122(B) apply to an adjustable speed drive system that uses a bypass device. This provision requires the ampacity rating of the branch-circuit conductors to be the larger of either of the following requirements: **(1)** 125 percent (1.25) of the rated input current to the power conversion equipment *or* **(2)** 125 percent (1.25) of the full-load current rating of the motor being served per NEC 430.6. The branch-circuit conductor requirements of NEC 430.122(B) also apply the same procedure as used for continuous loads, motor loads, motor-compressor loads and others.

Table 430.249 - Full-Load Current, Two-Phase Alternating-Current Motors (4-Wire)

54. How much current flows through the common conductor of a 2-phase, 60 horsepower motor rated for 230V operating on a 2-phase, 3- wire system?

Refer to the description found beneath the heading of Table 430.249. According to the second sentence of the paragraph of the description, the current in the common conductor of a 2-phase, 3-wire system will be 1.41 times the value given in Table 430.249.

For the motor mentioned, the full-load current is 133A. This is the amount of current that will flow through each phase. To determine the current flow through the common conductor the full-load current requires being increased 1.41 times which is,

$$133A \times 1.41 = 187.53A$$

The amount of current that will flow through the common conductor is 187.53A. The results of this calculation simply means that once a 3-wire, 2-phase motor is connected to a similar electrical system, the motor's common conductor would require being sized larger than the phase conductors.

Table 430.250 - Full-Load Current, Three-Phase Alternating-Current Motors

55. Two synchronous 30HP, 3 phase motors are rated 460 volts. The nameplate of one motor list a power factor rating of .80 while the other motor has a power factor rating of .90. Determine each motor's full-load current.

The *footnote* to Table 430.250 provides mandatory instructions for calculating the full-load current for synchronous motors having other than a unity power factor (100 percent). Because both motors apply, the *footnote* states that the full-load current for each motor must be determined by increasing the corresponding full-load current values in Table 430.250, 1.1 times for synchronous motors with a .90 (90 percent) power factor and 1.25 times for synchronous motors with a .80 (80 percent) power factor. Per Table 430.250 the full-load current for both motors is 32A. Applying the multiplier for each motor per power factor, the motor's full-load currents are now,

@ .90 power factor   @ .80 power factor

$$32A \times 1.1 = 35.2A \qquad 32A \times 1.25 = 40A$$

As it pertains to synchronous motors, these increased full-load current values much be used in accordance with NEC 430.6(A)(1) for determining the ampacity of conductors, the ampere ratings of switches and the overcurrent protection needed for these type motors.

Tables 430.251(A) and **430.251(B)** - Conversion Tables for Single- and Three-Phase Locked-Rotor Current

Refer to question Nos. 3., 19. and 20. of Article 440 and question No. 4. of Article 455 for applications relative to Tables 430.251(A) and 430.251(B).

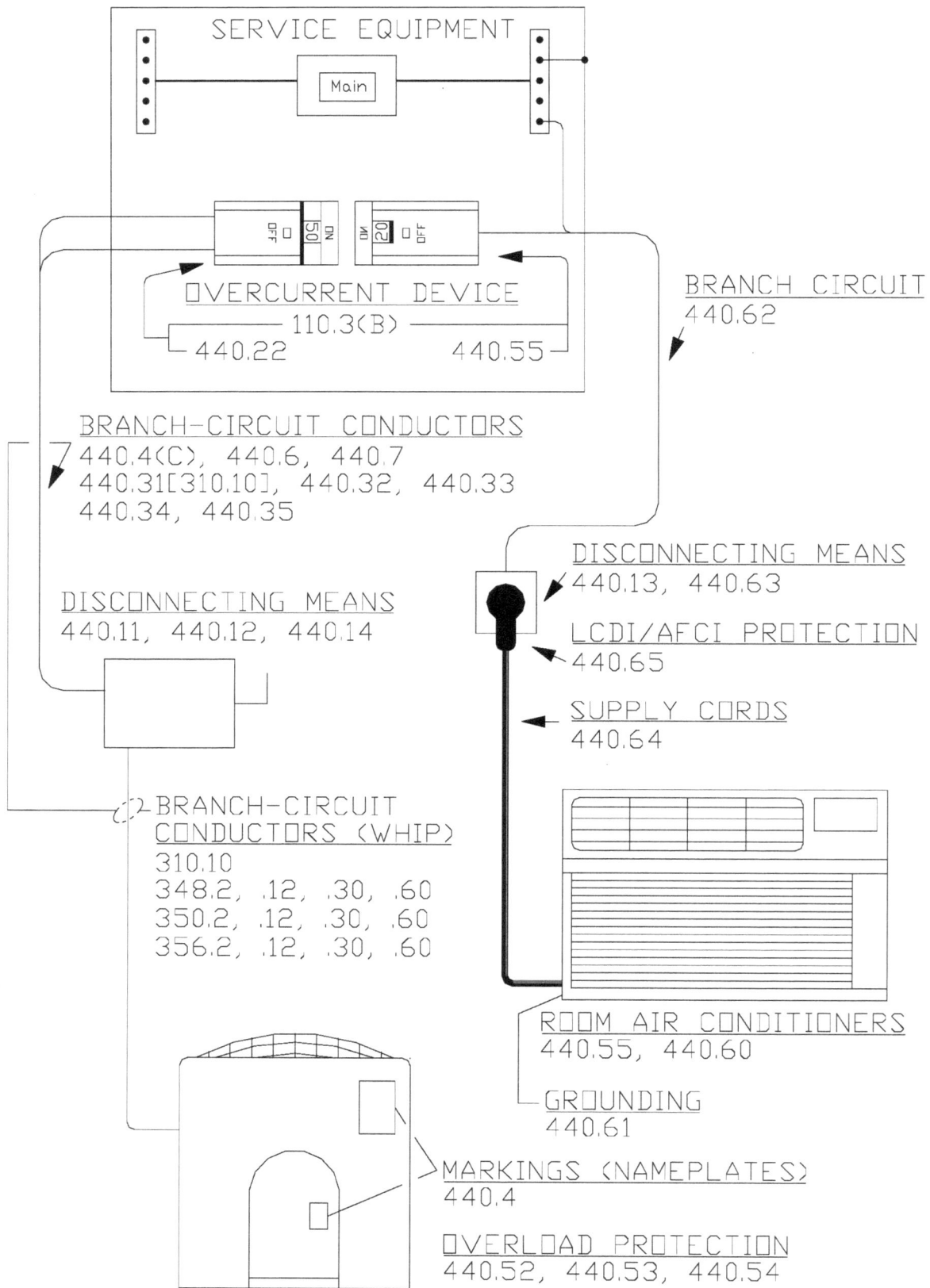

SERVICE EQUIPMENT

Main

OVERCURRENT DEVICE
110.3(B)
440.22   440.55

BRANCH CIRCUIT
440.62

BRANCH-CIRCUIT CONDUCTORS
440.4(C), 440.6, 440.7
440.31[310.10], 440.32, 440.33
440.34, 440.35

DISCONNECTING MEANS
440.13, 440.63

LCDI/AFCI PROTECTION
440.65

DISCONNECTING MEANS
440.11, 440.12, 440.14

SUPPLY CORDS
440.64

BRANCH-CIRCUIT
CONDUCTORS (WHIP)
310.10
348.2, .12, .30, .60
350.2, .12, .30, .60
356.2, .12, .30, .60

ROOM AIR CONDITIONERS
440.55, 440.60

GROUNDING
440.61

MARKINGS (NAMEPLATES)
440.4

OVERLOAD PROTECTION
440.52, 440.53, 440.54

ARTICLE 440 - Air-Conditioning and Refrigerating Equipment

Considered an extension of Article 430, Article 440 focus primarily on a specific type motor. It provides pertinent information that supports the electrical circuitry and operation of hermetically sealed motor-compressors. The term *hermetic* entails something that is totally enclosed, airtight and permanently sealed. As it pertains to air-conditioning and refrigeration equipment, that something is, the combination of a motor and compressor where both components are contained in the same enclosure which requires the motor to operates in a cooling substance called a refrigerant. Unlike a standard motor, hermetically sealed motor-compressors do not have an external shaft or shaft seals.

440.6(A) - Hermetic Refrigerant Motor-Compressor

For a hermetic-refrigerant motor-compressor, the rated-load (full-load) current marked on the nameplate of the air-conditioning unit enclosing the motor-compressor must be applied to determine the rating *or* ampacity of the disconnecting means, the branch-circuit conductors, the controller, the branch-circuit overcurrent protection *or* the rating for the overload protection. In the event the rated-load current is not marked on the nameplate of the air-conditioning unit the rated load-current shown on the name-plate of the motor-compressor must be used.

Where the branch-circuit selection current (bcsc) is marked on the nameplate of the air-conditioning unit it must be used instead of the rated-load current marked on the air-conditioning unit or the motor-compressor, if greater. The branch-circuit selection current must never be used for determining the overload protection for a hermetic-refrigerant motor-compressor, only the rated-load current must be applied.

440.7 - Highest Rated (Largest) Motor

Refer to Article 430 for standard size motors. See question No. 15. (NEC 430.17 - Highest Rated or Smallest Rated Motor). The same procedure is used for hermetic refrigerant motor-compressors where one of the applicable current ratings listed in NEC 440.6(A) is applied.

440.12(A) - Hermetic Refrigerant Motor-Compressor (Rating and Interrupting Capacity)

The disconnecting means serving a hermetic refrigerant motor-compressor must be sized based on the **(1)** nameplate rated-load [full-load] current *or* the branch-circuit selection current [if greater] of the air-conditioning unit enclosing the motor-compressor and the **(2)** motor-compressor's locked-rotor current, respectively of the motor-compressor per NEC 440.12(A)(1) and (2).

440.12(A)(1) - Ampere Rating (Hermetic Refrigerant Motor-Compressor)

NEC 440.12(A)(2) reference the use of NEC 430.109. NEC 430.109 lists the different type devices that can be used as disconnecting means for standard motors per Article 430 and hermetic refrigerant motor-compressors per Article 440. Per NEC 430.109, the disconnecting means may be a horsepower-rated or molded case switch, circuit breaker or controller.

Per Article 100 a *disconnecting means* is defined as a device, or group of devices, or other means by which the conductors of a circuit can be disconnected from their source of supply. Being disconnected from a source of supply does not altogether imply that a means of overcurrent protection is inclusive of such a device or a group of devices. Therefore, regardless of whether the means of overcurrent protection is inclusive of such device(s) or located upstream to a disconnecting means, the need for overcurrent protection must adhere to the provisions of NEC 440.22(A) where a single hermetic refrigerant motor-compressor is involved and NEC 440.22(B) where hermetic refrigerant motor-compressors along with other type loads are engaged. A disconnecting means alone does not necessarily suggest the inclusion of overcurrent protection.

1. What size circuit breaker is required to serves as the disconnecting means for a hermetic refrigerant motor-compressor if the nameplate of an A/C unit enclosing the motor-compressor displays a rated-load current of 36A and a branch-circuit selection current of 43.7A?

Per NEC 440.12(A)(1) the ampere rating of the disconnecting means where in this case is a circuit breaker requires the rating of the circuit breaker to be sized no less than 115 percent (1.15) of the larger name-plate current. Because the branch-circuit selection current is the larger of the two nameplate currents it must be used to size the circuit breaker. As a result,

$$43.7A \times 1.15 = 50.26A$$

and a 60A circuit breaker is required. A 50A circuit breaker could not be used because the 115 percent increase would be violated (50A/43.7 = 1.14 rounded off).

440.12(A)(2) - Equivalent Horsepower (Hermetic Refrigerant Motor-Compressor)

2. Assume the hermetic refrigerant motor-compressor in question No. 1. is rated for 3-phase at 480V. If the motor-compressor displayed a nameplate locked-rotor current of 213A what size horsepower rated disconnecting means would be required?

Per NEC 440.12(A)(2), to determine the equivalent horsepower in complying with NEC 430.109, the horsepower rating shall be selected from Tables 430.248 - .250 which corresponds to the nameplate rated-load current *or* branch-circuit selection current, whichever is greater along with the horsepower rating from either Table 430.251(A) or 430.251(B) which corresponds to the locked-rotor current. Where the nameplate rated-load current *or* branch-circuit selection current and locked-rotor current do not correspond to the currents shown in the given tables, the horsepower rating corresponding to the next higher value shall be selected.

Because this is a 3-phase motor compressor, Table 430.250 will be used to gather a corresponding horsepower rating per nameplate branch-circuit selection current which is greater than the rated-load current while Table 430.251(B) will be used to gather a corresponding horsepower rating per nameplate locked-rotor current.

Per Table 430.250 at 460V 3-phase, a corresponding horsepower rating per branch-circuit selection current at 43.7A would be equivalent to the next higher rated motor which is a 40-horsepower motor (52A).

Per Table 430.251(B) at 480V 3-phase, a corresponding horsepower rating per locked-rated current at 213A would be equivalent to the next higher rated motor which is a 30-horsepower motor (218A).

In comparison of the two selected horsepower ratings, the horsepower rated disconnecting means must have a rating of 40-horsepower based on the corresponding branch-circuit selection current.

440.12(B)(1) - Horsepower Rating (Combination Loads)

3. Refer to Figure 440.12(B)(1)-3. Determine the size disconnecting means needed based on the given loads. All loads have the potential of operating at the same time.

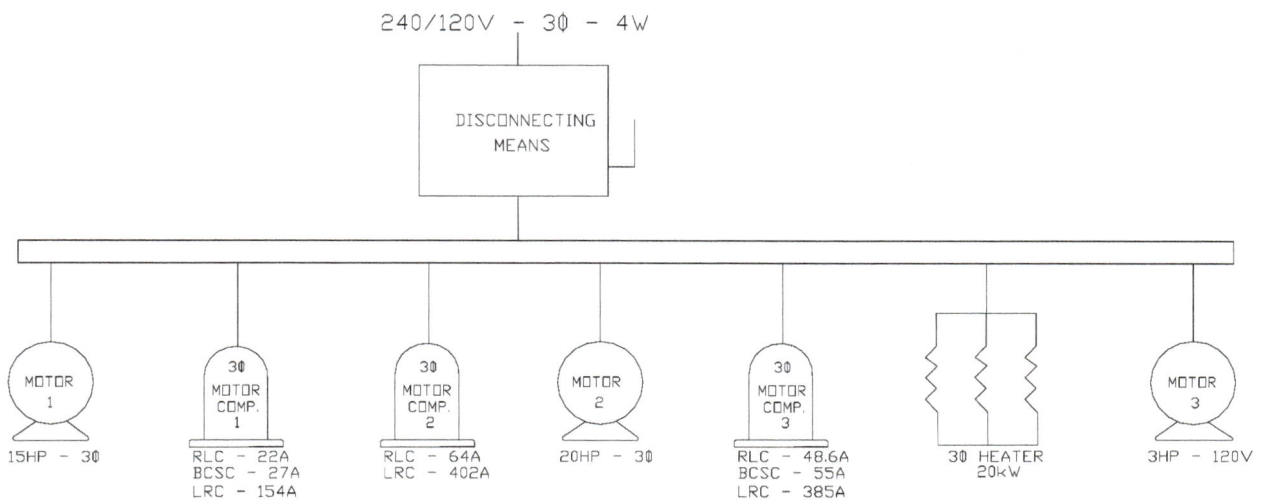

Figure 440.12(B)(1)-3

NEC 440.12(B) states when two or more hermetic refrigerant motor-compressors *or* one or more hermetic refrigerant motor-compressors with other motors *or* loads may be simultaneous on a single disconnecting means, the rating for the disconnecting means shall be determined in accordance with 440.12(B)(1) and (B)(2). The disconnecting means must be adequately sized based on both horsepower rating and ampere rating.

Where a disconnecting means is determined based on horsepower rating, NEC 440.12(B)(1) provide such guidelines. Per NEC 440.12(B)(1), to determine the required horsepower rating of a disconnecting means all currents, which includes resistance load currents, rated-load (full-load) currents and locked rotor currents are considered.

The combined rated-load and locked-rotor current of both motor and motor-compressor loads must be considered as a single motor to meet the requirements of subsections (a) and (b).

Applying the provisions of **NEC 440.12(B)(1)(a)**.

Per NEC 440.12(B)(1)(a)
Using the information provided in Figure 440.12(B)(1)-3.

Full-load currents [Tables 430.250 and .248]	**RLC**	**BCSC** [Figure 440.12(B)(1)-3]	****Heater** [other load]
Motor 1 (15HP - 3φ) - 42A	*MC 1 -	27A	48.11A
Motor 2 (20HP - 3φ) - 54A	MC 2 64A	-	
Motor 3 (3HP - 1φ) - 34A	MC 3 -	55A	

* MC = Motor-Compressor

** The 20kW 3-phase heater load's rated-load current per calculation is 20kW/240V x √3 = 48.11A.

The equivalent full-load current for the combined load as calculated amounts to the following:

$$42A + 54A + 34A + 27A + 64A + 55A + 48.11A = \mathbf{324.11A}$$

Applying the provisions of **NEC 440.12(B)(1)(b)**.

Per NEC 440.12(B)(1)(b)

Using the information provided in Figure 440.12(B)(1)-3.

Locked-rotor currents [Tables 430.251(A) and (B)]	**LRC** [Figure 440.12(B)(1)-3]	**Heater** [other load]
Motor 1 (15HP - 3φ) - 232A	*MC 1 154A	48.11A
Motor 2 (20HP - 3φ) - 290A	MC 2 402A	
Motor 3 (3HP - 1φ) - 204A	MC 3 385A	

* MC = Motor-Compressor

The equivalent locked-rotor current for the combined load as calculated amounts to the following:

$$232A + 290A + 204A + 154A + 402A + 385A + 48.11A = \mathbf{1715.11A}$$

In accordance with **NEC 440.12(B)(1)(a)** the horsepower rating corresponding to the value exceeding **324.11A** is selected *per Table 430.250* resulting to a 150-horsepower rated disconnecting means at 240V 3-phase.

In comparison to **NEC 440.12(B)(1)(b)**, the horsepower rating corresponding to the value exceeding **1715.11A** is selected *per Table 430.251(B)* resulting to a 125-horspower rated disconnecting means at 240V 3-phase.

Where applied, the disconnecting means based on the higher horsepower rating (150-horsepower) is selected.

In applying the applications of NEC 440.12(B)(1)(a) and (b), the following formulas can be used for calculating the combined current values:

Formula per 440.12(B)(1)(a)

[motor(s) full-load current(s) per Tables 430.248 - 430.250] + [*motor-compressor(s) rated-load current(s) or branch-circuit selection current(s), whichever is greater*] + [rated-load current(s) (ampere ratings) of other loads] = equivalent full-load current for the combined load

Formula per 440.12(B)(1)(b)

[motor(s) locked-rotor current(s) per Tables 430.251(A) and/or (B)] + [*motor-compressor(s) locked-rotor current(s)*] + [rated-load current(s) (ampere ratings) of other loads] = equivalent locked-rotor current for the combined load.

440.12(B)(2) - Full-Load Current Equivalent (Combination Loads)

4. Based on the equivalent full-load current calculated in question No. 3. determine the ampere rating of the disconnecting means required.

Per NEC 440.(B)(2) the ampere rating of the disconnecting means shall be at least 115 percent (1.15) of the sum of all currents at the rated-load condition. As a result,

$$324.11A \times 1.15 = 372.73A$$

The disconnecting means as a minimum must be rated for 372.73A. In summary, the disconnecting means used for the installation in Figure 440.12(B)(1)-3 must carry an ampere rating no less than 372.73A at 150HP. However, if the disconnect switch in Figure 440.12(B)(1)-3 was an unfused motor circuit switch, without fuse holders the *Exception* to NEC 440.12(B)(2) states an ampere rating less than 372.73A could be used bearing the 150HP rating is still applied.

440.12(C) - Small Motor-Compressors (Rating and Interrupting Capacity)

See question No. 53. of Article 430 [430.110(C)(3)] for same application.

440.22(A) - Rating or Setting for Individual Motor-Compressor (Application and Selection)
     (5. - 7.) [3]

5. The nameplate of an A/C unit that's equipped with a motor-compressor list the unit's rated current at 37A and the branch-circuit current at 40.6A. What size overcurrent protection is required to protect the unit?

Per NEC 440.22(A), the rating or setting of the overcurrent (short-circuit and ground-fault) protective device for an individual motor-compressor must not exceed 175 percent (1.75) of the motor-compressor rated-load *or* branch-circuit selection current, whichever is greater. Because the branch-circuit selection current is the larger of the two given nameplate currents, it is used to size the needed overcurrent protection for the A/C unit. Based on the branch-circuit selection current and the allowed percent multiplier, the following results are provided:

$$40.6A \times 1.75 = 71.05A$$

Because the selection of the overcurrent protection is limited to 175 percent, the next standard size overcurrent device (80A) per NEC 240.6(A) is not allowed based on the calculated results. As a *minimum*, a **70A** overcurrent protective device is required.

6. If the 70A overcurrent protective device in question No. 5. will not allow the A/C unit to start, what size device is permitted?

NEC 440.22(A) permits the rating or setting of such overcurrent protective device to be sized up to 225 percent (2.25) when the protection specified is not sufficient for the starting current of the motor-compressor. Again, based on the branch-circuit selection current and the allowed increased percent multiplier, the following results are provided:

$$40.6A \times 2.25 = 91.35A$$

Because the selection of the overcurrent protection is limited to 225 percent, the next standard size overcurrent device (100A) per NEC 240.6(A) is not allowed based on the calculated results. As a *maximum* rating, a **90A** overcurrent protective device is permitted.

7. Determine the minimum and maximum overcurrent protective devices needed for each individual load in Figure 440.12(B)(1)-3. Assume the use of an inverse-time circuit breaker for the standard motor loads.

Standard Motors overcurrent protection

Minimum - Per Table 430.52 (single-phase and AC polyphase motors) the overcurrent protection devices for the motor loads are based on 250 percent (2.50) of the motor's full-load current where inverse time circuit breakers are used. *Exception No. 1* to NEC 430.52(C)(1) permits the next standard size overcurrent device to be used.

Maximum - *Exception No. 2(c)* to NEC 430.52(C)(1) permits the rating of an inverse time circuit breaker to be increased up to 400 percent (4) when the motor's full-load current does not exceed 100 amperes.

Motor-compressors overcurrent protection

Minimum - Per NEC 440.22(A), the overcurrent protection device must have a rating or setting not exceeding 175 percent (1.75) of the motor-compressor rated-load current *or* branch-circuit selection current, whichever is greater.

Maximum - Per NEC 440.22(A), where the protection specified is not sufficient for the starting current of the motor-compressor's, the rating or settingshall be permitted to be increased but not exceed 225 percent (2.25) of the motor-compressor's rated-load current *or* branch-circuit selection current, whichever is greater.

Heater overcurrent protection

Per NEC 424.3, fixed electric space-heating equipment shall be considered a continuous load. Beacause the heater load is considered continuous, the overcurrent protection device must be rated at 125 percent (1.25) of the heater load. There are no minimum and maximum ratings for the device protecting this load. NEC 240.4(B) permits the next standard size overcurrent device to be used.

Based on the given provisions the overcurrent protection for each load is as follow:

		Minimum OCD			Maximum OCD
Motor 1 - 42A x 2.50 =	105A	110A	42A x 4 = 168A	150A
Motor 2 - 54A x 2.50 =	135A	150A	54A x 4 = 216A	200A
Motor 3 - 34A x 2.50 =	85A	90A	34A x 4 = 136A	125A
MC 1 - 27A x 1.75 =	47.25A	45A	27A x 2.25 = 60.75A	60A
MC2 - 64A x 1.75 =	112A	110A	64A x 2.25 = 144A	125A
MC3 - 55A x 1.75 =	96.25A	90A	55A x 2.25 = 123.75	110A
Heater - 48.11A x 1.25 = 60.14A		70A		--

440.22(B) - Rating or Setting for Equipment (Application and Selection)

Refer to 440.22(B)(1) and (2).

440.22(B)(1) - Motor-Compressor Largest Load (Application and Selection) (8. - 10.) [3]

8. What size overcurrent protective device is required to protect a feeder circuit supplying three motor-compressors rated for 62A, 48.5A, and 36.4A?

Where two or more hermetic refrigerant motor-compressor loads are on the same circuit, the overcurrent protective device protecting such loads is required to be sized at 175 percent of the largest motor-compressor's rated-load current *or* branch-circuit selection current per NEC 440.22(A) *plus* the sum of the rated-load current(s) *or* branch-circuit selection current(s) of all other motor-compressor(s) and the ratings of all other motor and non-motor loads supplied.

Applying the given provisions, the required overcurrent protective device can be determined based on the following calculations where the motor-compressor rated for 62A is the largest of all other motor-compressors.

$$62A \times 1.75 \, (108.5A) + 48.5A + 36.4A = 193.4A$$

Because NEC 440.22(B)(1) states that the (circuit's) protective device shall not exceed *the value* (108.5A, not the rating or setting of a protective device) specified in 440.22(A) for *the largest motor-compressor plus* the sum of the larger rated-load current *or* branch-circuit selection current of all other motor-compressor(s), the next standard size overcurrent protective device (200A) is not allowed, refer to NEC 240.6(A). As a result, a 175A overcurrent device is required.

9. If the 175A overcurrent protective device in question No. 8. causes nuisance tripping of the circuit, what size device is permitted to prevent such occurrences?

Based on all the given information provided in question No. 8., the overcurrent protective device is allowed to be sized up to 225 percent (2.25) per NEC 440.22(A) using the same procedure provided in NEC 440.22(B)(1). Therefore,

$$62A \times 2.25\ (139.5A) + 48.5A + 36.4A = 224.4A$$

Again, because NEC 440.22(B)(1) states that the (circuit's) protective device shall not exceed *the value* specified in 440.22(A) for *the largest motor-compressor plus* the sum of the larger rated-load current *or* branch-circuit selection current of all other motor-compressor(s), the next standard size overcurrent protective device (225A) is not allowed, refer to NEC 240.6(A). As a result, a 200A overcurrent protective device is only allowed.

10. If the disconnecting means in Figure 440.12(B)(1)-3 was a fusible switch, what size dual element-time delay fuses would be required to serve as the overcurrent protection for this installation? Consider both minimum and maximum sizes.

Motor-compressor No. 2 is the largest load connected to the circuit based on a rated-load current of 64A. Applying the same procedures as in question Nos. 8. and 9.

$$64A \times 1.75\ (112A) + 42A + 27A + 54A + 55A + 48.11 + 34A = 372.11A$$

Because NEC 440.22(B)(1) states that the protective device shall not exceed the value specified in 440.22(A) for the largest motor-compressor *plus* the sum of the larger rated-load current *or* branch-circuit selection current of all other motor-compressor(s), the next standard size overcurrent protective device (400A) is not allowed, refer to NEC 240.6(A). As a minimum, 350A dual element-time delay fuses are required.

As for the maximum overcurrent protection.

$$64A \times 2.25\ (144A) + 42A + 27A + 54A + 55A + 48.11 + 34A = 404.11A$$

Again, because NEC 440.22(B)(1) states that the protective device shall not exceed the value specified in 440.22(A) for the largest motor-compressor *plus* the sum of the larger rated-load current *or* branch-circuit selection current of all other motor-compressor(s), the next standard size overcurrent protective device (450A) is not allowed, refer to NEC 240.6(A). As a maximum, 400A dual element-time delay fuses are required.

440.22(B)(2) - Motor-Compressor Not Largest Load (Application and Selection)

11. A 460V, three-phase feeder supplies two hermetic motors and three standard (non-hermetic) motors to support the air-conditioning load of a four-story office building. If the hermetic motors are rated for 65A and 74A and the standards motors are rated for 30HP, 60HP, and 75HP, what size overcurrent protective device is required to protect the feeder supplying these motor loads?

Based on the given horsepower ratings, the full-load current of each standard motor is 40A, 77A, and 96A respectively per Table 430.250.

In an application such as this where a motor-compressor is not the largest load, the provisions of NEC 440.22(B)(2) are applied along with NEC 430.53(C)(4) as referenced. Combined, NEC 440.22(B)(2) and 430.53(C)(4), in simple terms states when motor-compressors and standard motors are installed on the same circuit and a hermetic refrigerant motor-compressor is not the largest of the motor loads, similar procedures as outlined in NEC 430.62 are followed. NEC 430.53(C)(4) requires the overcurrent protective device for an application such as this, be it fuses or inverse time circuit breakers to have a rating *not exceeding that specified in Table 430.52* for the highest rated motor *plus* the sum of the ratings/full-load current ratings of all remaining hermetic/standard motors being supplied.

For this application, dual element time-delay fuses and an inverse time circuit breaker will both be considered. Referring to Table 430.52, for three-phase motors, 175 percent (1.75) and 250 percent (2.50) of the largest motor's full-load current must be applied respectively for sizing dual element time-delay fuses and an inverse time circuit breaker.

Applying the full-load current of the largest non-hermetic motor and the given percentage values where,

$$96A \times 1.75 = 168A \text{ (dual element time-delay fuses)} = 150A \text{ fuses}$$
and
$$96A \times 2.50 = 240A \text{ (inverse time circuit breaker)} = 225A$$

Now, if dual element time-time fuses are used to protect the circuit, the rating of the fuses is determined based upon the total of all current values;

$$168A + 65A + 74A + 40A + 77A = 424A$$

where the rating of the fuses is limited to 400A because the provisions states, "shall not exceed" a value equal to the summation of all involved current values.

In conclusion, if an inverse time circuit breaker is used to protect the circuit, the rating of the breaker is determined in the same manner as the fuses;

$$240A + 65A + 74A + 40A + 77A = 496A$$

where the rating of the circuit breaker is limited to 450A based upon the same provisions as stated for the dual element time-time fuses.

Although the deciding ratings of the dual element time-delay fuses (400A) and the inverse time circuit breaker (450A) fell short of the total current values (424A and 496A); both overcurrent devices still exceed the actual load current 352A (65A + 74A + 40A + 77A + 96A).

440.32 - Single Motor-Compressor (12. - 14.) [3]

12. The nameplate current for a 3-phase, 440V, 10-ton hermetically sealed refrigerant motor-compressor is 43.6 amps. What size THWN copper branch-circuit conductors are required to supply the motor-compressor?

NEC 440.32 requires the branch-circuit conductors to a single motor-compressor to have an ampacity not less than 125 percent (1.25) of the motor-compressor rated-load current. Based on the given motor-compressor current where,

$$43.6A \times 1.25 = 54.5A$$

The branch-circuit conductors must have a minimum rated ampacity of 54.5A. Per Table 310.15(B)(16), 6 AWG THWN copper conductors which have a rated ampacity of 65A are required.

13. If the nameplate of the motor-compressor unit in question No. 12. displayed a branch-circuit selection current of 47.3A, could the same size branch-circuit conductors be used?

In its entirety, NEC 440.32 requires the larger of the rated-load current *or* the branch-circuit selection current of a motor-compressor to be used when sizing the branch-circuit conductors of a single motor-compressor. Because the given 47.3A branch-circuit selection current is larger than the 43.6A nameplate (rated) current, the branch-circuit conductors must be sized accordingly. Therefore,

$$47.3A \times 1.25 = 59.13A$$

Based on the calculated ampacity at 59.13A, the same size (6 AWG THWN copper) branch-circuit conductors can be used.

14. Determine the minimum conductor ampacity needed to supply each individual load in Figure 440.12(B)(1)-3.

According to NEC 430.22, 424.3(B) and 440.32, the conductors supplying each given load requires a 125 percent (1.25) increase of a respective full-load current, continuous load, and either the greater of a rated-load current *or* a branch-circuit selection current.

Based on the referenced information, the following calculations are as follows:

		Minimum Conductor Ampacity
Motor 1 - 42A x 1.25	=	52.5A
Motor 2 - 54A x 1.25	=	67.5A
Motor 3 - 34A x 1.25	=	42.5A
MC 1 - 27A x 1.25	=	33.75A
MC2 - 64A x 1.25	=	80A
MC3 - 55A x 1.25	=	68.75A
Heater - 48.11A x 1.25	=	60.14A

For wye-start, delta-run connected motor-compressor branch-circuit conductors, refer to question No. 21. of Article 430 for a similar calculation.

440.33 - Motor-Compressor(s) With or Without Additional Motor Loads (15. - 17.) [3]

15. With a fan motor that pulls 2.1A, what size TW copper branch-circuit conductors are required to supply a single-phase, 240V, hermetic motor-compressor that's rated for 29.8A?

In general, NEC 440.33 requires conductors supplying one or more motor-compressor(s) with or without additional motor loads to have an ampacity not less than 125 percent of the highest rated motor or motor-compressor* *plus* the full-load current ratings of all other motors of the group.

Based on the motor-compressor having the highest rating of the two motors the resulting calculation amounts to the following,

$$29.8A \times 1.25 + 2.1A = 39.35A$$

At 39.35A, 8 AWG TW copper conductors which have a rated ampacity of 40A per Table 310.15(B)(16) are required.

*25 percent of the highest motor or motor-compressor rating based on the larger selected current rating *plus* 100 percent of the larger selected current rating.

16. Refer to question No. 11. What size feeder conductors are required to supply the motors? Consider the use of both THWN copper and aluminum conductors.

Based on the given horsepower ratings the full-load currents of each standard motor are per Table 430.250 are 40A, 77A, 96A respectively.

In compliance with the provisions provided in NEC 440.33, the rated load of each motor-compressor (hermetic motors) *plus* the full-load currents of all other motors *plus* 25 percent (.25) of the highest motor or motor-compressor rating in the group will be followed as outlined. In this particular application, the 75HP motor which has a 96A full-load current is considered the highest motor. Therefore,

$$65A + 74A + 40A + 77A + 96A + (96A \times .25) = 376A$$

Per Table 310.15(B)(16), either 500 kcmil copper (380A) or 750 kcmil aluminum (385A) THWN conductors can be used as the feeder conductors to supply the motor loads.

17. Determine the minimum conductor ampacity needed to supply the loads in Figure 440.12(B)(1)-3.

Applying the same procedure as the previous question,

$$42A + 27A + 64A + 54A + 55A + 48.11A + 34A + 64A \times .25 \ (16A) = 340.11A$$
$$or$$
$$64A \times 1.25 \ (80A) + 42A + 27A + 54A + 55A + 48.11A + 34A = 340.11A$$

Both procedures produce the same results. A minimum conductor ampacity of 340.11A is needed to supply the loads.

440.34 - Combination Load

18. Two 16A motor-compressors along with a 15A appliance load and an 18A lighting and receptacle load are supplied by the same conductors. Determine the minimum required ampacity of the supplying conductors.

NEC 440.34 requires conductors supplying a motor-compressor load in addition to other load(s) (lighting [and]or appliance, etc.) to have an ampacity not less than 125 percent (.125) of the highest rated motor-compressor current per NEC 440.32 or 440.33 *plus* other load(s) (lighting (and)or appliance, etc.). As a result, 16A (highest rated motor-compressor current) x 1.25 + 16A (other motor-compressor current) + 15A (appliance load) + 18A (lighting and receptacle load) = 69A

Based on the calculated load, the minimum ampacity of the supplying conductors must be 69A.

440.41(A) - Motor-Compressor Controller (Rating)

19. If motor starters were used to control the motor-compressors in Figure 440.12(B)(1)-3 what sizes would be required?

Similar to the procedures used in NEC 440.12(A) and (B)(1) to size the disconnecting means in Figure 440.12(B)(1)-3, motor-compressor controller are sized in the same manner per NEC 440.41(A). NEC 440.41(A) states that a motor-compressor controller shall have both a continuous-duty full-load current rating and a locked-rotor current rating not less than the nameplate rated-load current *or* branch-circuit selection current, whichever is greater and locked-rotor current, respectively, of the compressor. In the event the motor controller is horsepower rated but is without one or both of the foregoing current ratings, an equivalent full-load current shall be determined per Tables 430.248 - 430.250 and Tables 430.251(A) and (B) for an equivalent locked-rotor current.

Based on the motor-compressors' given currents the required starters are sized accordingly:

	Rated-load current *or* Branch-circuit selection current		Equivalent Horsepower Rating per Tables 430.250
MC 1 -	27A	=	10
MC 2 -	64A	=	25
MC 3 -	55A	=	25

	Rated locked-rotor current		Equivalent Horsepower Rating per Tables 430.251(B)
MC 1 -	154A	=	10
MC 2 -	402A	=	30
MC 3	385A	=	30

Based on the calculated results where the larger of both horsepower ratings is applied in sizing the required starters for each motor-compressor, for motor-compressor MC1 a 10-horsepower rated starter is required whereas for motor-compressors MC2 and MC3, 30-horsepower rated starters are required.

440.41(B) - Controller Serving More Than One Load (Rating)

20. If the motor-compressors in Figure 440.12(B)(1)-3 were required to be controlled by one starter, what size starter would be needed?

NEC 440.41(B) states where a controller serves more than one load, the combined load must be determined in accordance with NEC 440.12(B) in sizing a single controller.

Applying the total current values of the three motor-compressors:

Rated-load currents *or* Branch-circuit selection currents		Equivalent Horsepower Rating per Tables 430.250
MC 1, 2 and 3 - 27A + 64A + 55A = 146A	=	60

Rated locked-rotor current		Equivalent Horsepower Rating per Tables 430.251(B)
MC 1, 2 and 3 - 154A + 402A + 385A = 941A	=	75

Based on the calculated results, where the larger of both horsepower ratings is applied in sizing the required starter for the motor-compressor loads, a 75-horsepower rated starter is required.

440.52(A) - Protection of Motor-Compressor (Application and Selection)

21. What size overload relays are needed to protect the motor-compressors in Figure 440.12(B)(1)-3?

In accordance with NEC 440.52(A)(1) a separate overload relay shall be selected to trip at not more than 140 percent (1.40) of the motor-compressor rated load-load current.

Applying the rated-load current of each motor-compressor, the results are as follows:

	Rated-load current		Rating of overload relay
MC 1 -	22A x 1.40	=	30.8A
MC 2 -	64A x 1.40	=	89.6A
MC 3 -	48.6A x 1.40	=	68.04A

If overload protection was required for the motors in Figure 440.12(B)(1)-3, the provisions of NEC 430.32(A)(1) and (C) must be applied.

For questions pertaining to motor overload protection refer to question Nos. 31. - 36. of Article 430.

440.62(B) - Where No Other Loads Are Supplied (Branch-Circuit Requirements)

22. The nameplate current of a window air-conditioning unit is rated for 12.7A. If the unit is cord-and-plug connected and supplied by an individual branch-circuit, what size overcurrent device is required to protect the unit?

According to NEC 440.62(B), where no other loads are supplied, the rating of a branch-circuit must not exceed 80 percent (.80) of the total marked rating of a cord-and-attachment-plug connected room air-conditioner.

Because the rating of a branch-circuit is determined by the overcurrent device protecting the circuit, if a 15A overcurrent device was used it would be inadequate because the rating of the air-conditioner would exceed 80 percent of its rating (15A x .80 = 12.5A). Therefore, a 20A overcurrent device would be required (20A x .80 = 16A, where 16A exceeds the 12.7A rating of the air-conditioner).

440.62(C) - Where Lighting Units or Other Appliances Are Also Supplied (Branch-Circuit Requirements)

23. If a 15A overcurrent device was used to protect a circuit supplying a room air-conditioner along with other lighting and receptacle outlets how much load could the air-conditioner place on the circuit?

According to NEC 440.62(C), when a branch circuit supplies lighting and receptacle outlets along with a cord-and-plug connected room air-conditioner, only 50 percent (.50) of the circuit's rating can be used towards supplying the marked rating of the air-conditioner. As a result, if placed on a 15A circuit, the marked rating of the air-conditioner must be rated for 7.5A or less (15A x .50 = 7.5A).

ARTICLE 445 - Generators

445.13 - Ampacity of Conductors

1. A 3ϕ, 45kVA generator rated for 240/120V has a 90 percent power factor. The generator's nameplate current rating is 156 amps. The generator will be used to supply a 150A emergency lighting panelboard. What size copper conductors rated for 75°C are needed to feed the panelboard?

According to NEC 445.13 the ampacity of the conductors from the generator terminals to the first distribution device(s) shall not be less than 115 percent of the nameplate current rating of the generator. Considering this requirement,

$$156 \text{ amps x } 1.15 = 179.4 \text{ amps}$$

Referring to Table 310.15(B)(16), 3/0 AWG conductors rated for 75°C are needed to supply the calculated ampacity of 179.4 amps.

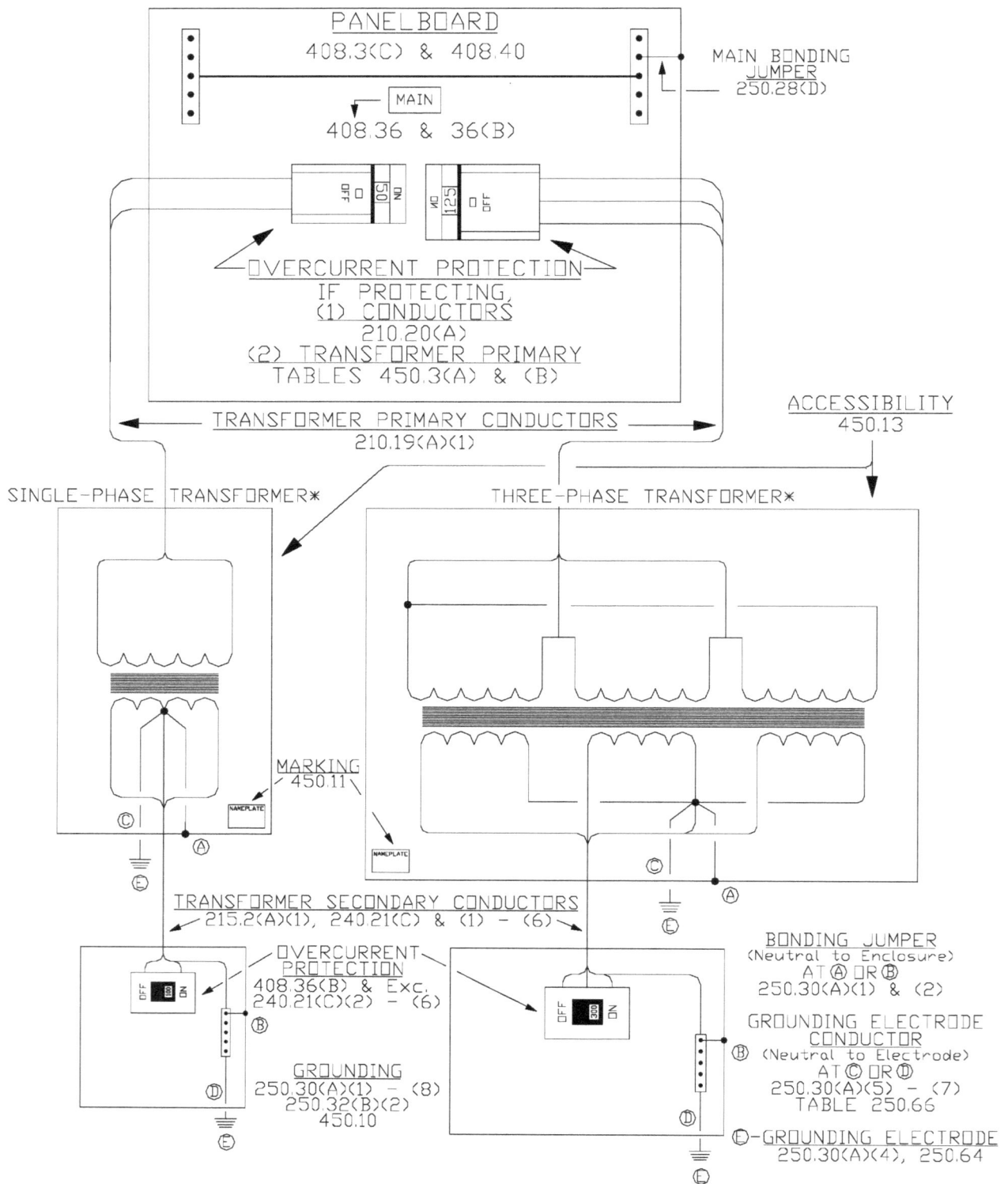

PANELBOARD
408.3(C) & 408.40

MAIN BONDING JUMPER
250.28(D)

MAIN
408.36 & 36(B)

OVERCURRENT PROTECTION
IF PROTECTING,
(1) CONDUCTORS
210.20(A)
(2) TRANSFORMER PRIMARY
TABLES 450.3(A) & (B)

TRANSFORMER PRIMARY CONDUCTORS
210.19(A)(1)

ACCESSIBILITY
450.13

SINGLE-PHASE TRANSFORMER*

THREE-PHASE TRANSFORMER*

MARKING
450.11

NAMEPLATE

NAMEPLATE

TRANSFORMER SECONDARY CONDUCTORS
215.2(A)(1), 240.21(C) & (1) - (6)

OVERCURRENT PROTECTION
408.36(B) & Exc.
240.21(C)(2) - (6)

BONDING JUMPER
(Neutral to Enclosure)
AT Ⓐ OR Ⓑ
250.30(A)(1) & (2)

GROUNDING ELECTRODE
CONDUCTOR
(Neutral to Electrode)
AT Ⓒ OR Ⓓ
250.30(A)(5) - (7)
TABLE 250.66

GROUNDING
250.30(A)(1) - (8)
250.32(B)(2)
450.10

Ⓔ-GROUNDING ELECTRODE
250.30(A)(4), 250.64

*SEPARATELY DERIVED EQUIPMENT - SEE DEFINITIONS - ARTICLE 100

ARTICLE 450 - Transformers and Transformer Vaults

The application of Article 450 <u>only</u> pertains to the actual transformer and the transformer's windings. Where supply and secondary conductors relative to transformers are involved, the application of Article 240 is applied. See discussion "Transformer Secondary Tap Conductors," Article 240 (Volume 2).

Any Location

Table 450.3(A) - Maximum Rating or Setting of Overcurrent Protection for Transformers Over 600 Volts (as a Percentage of Transformer-Rated Current) (1. - 4.) [4]

1. A 300kVA three-phase transformer is rated for 12.47kV-4160/2400V. The transformer's rated impedance is 3.7 percent. Determine the transformer's primary and secondary overcurrent protection if either circuit breaker(s) or fuses were used on both sides of the transformer.

Transformer rated impedance - A transformer's rated impedance is expressed as a voltage percentage. It is the percentage of rated primarily voltage that must be applied to cause rated secondary current to flow into a short-circuit at the secondary terminals. For example, a three-phase 12,470/465V, 1500kVA transformer with a 6.25 percent impedance would require only 779.38V (12,470V x .0625) of the primary voltage to produce the rated secondary current (1862.47A). One hundred (100) percent of the rated primary voltage will increase the rated secondary current 16 times (100/6.25) without the consideration of other factors to potentially produce a bolted (three-phase) fault current of 29,799.6A [(1500kVA/465V x 1.732) x 16].

Calculate transformer's primary and secondary full-load currents

$$I_P = \frac{300kVA\ (300,000VA)}{12.47kV(12,470V)\ x\ 1.732} = 13.89A$$

$$I_S = \frac{300kVA\ (300,000VA)}{4160V\ x\ 1.732} = 41.64A$$

Refer to Table 450.3(A)

Location: Any (unsupervised)
Impedance: Less than 6%
Primary and Secondary Voltage: Over 600 volts

Consider the required overcurrent protection

Table 450.3(A) **Note 1**, allows the next higher standard rated overcurrent device to be used when the current rating or setting does not correspond to the standard ampere ratings of NEC 240.6(A).

Primary Protection

If circuit breaker used [600% (6)]
13.89A x 6 = 83.34A. Use a 90A circuit breaker.

If fuses used [300% (3)]
13.89A x 3 = 41.67A. Use 45A fuses

Secondary Protection

If circuit breaker used [300% (3)]
41.64A x 3 = 124.92A. Use a 125A circuit breaker.

If fuses used [250% (2.5)]
41.64A x 2.5 = 104.1A. Use 110A fuses.

2. In question No. 1., if the transformers secondary voltage was rated for 480 volts what size circuit breaker or fuses would be required?

Calculate transformer's secondary full-load current

$$I_P = \frac{300kVA\ (300,000VA)}{480V\ x\ 1.732} = 360.85A$$

Secondary Protection

If circuit breaker or fuses used [125% (1.25)]
360.85A x 1.25 = 451.1A.

Use either a 500A circuit breaker or a set of fuses rated for 500A.

3. The nameplate of a transformer indicates the following:

kVA - 1000
Primary - 7200V
Secondary - 2400V
Phases - 3
% Z - 8.40V

Determine the transformer's primary and secondary overcurrent protection if either a circuit breaker or a set of fuses were used on both sides of the transformer.

Calculate transformer's primary and secondary full-load currents

$$I_P = \frac{1000kVA\ (1,000,000VA)}{7200V\ x\ 1.732} = 80.2A$$

$$I_S = \frac{1000kVA\ (1,000,000VA)}{2400V\ x\ 1.732} = 240.57A$$

Refer to Table 450.3(A)

Location: Any (unsupervised)
Impedance: Less than 6% not over 10%
Primary and Secondary Voltage: Over 600 volts

Consider the required overcurrent protection

Primary Protection

If circuit breaker used [400% (4)]
80.2A x 4 = 320.8A. Use a 350A circuit breaker.

If fuses used [300% (3)]
80.2A x 3 = 240.6A. Use 250A fuses.

Secondary Protection

If circuit breaker used [250% (2.5)]
240.57A x 2.5 = 601.43A. Use a 700A circuit breaker.

If fuses used [225% (2.25)]
240.57A x 2.25 = 541.28A. Use 600A fuses.

4. In question No. 3., if the transformer's secondary voltage was rated for 480 volts, what size circuit breaker or fuses would be required.

Calculate transformer's secondary full-load current

$$I_S = \frac{1000kVA\ (1,000,000VA)}{480V \times 1.732} = 1202.84A$$

Secondary Protection

If circuit breaker or fuses used [125% (1.25)]
1202.84A x 1.25 = 1503.55A. Use 1600A circuit breaker or fuses

Supervised Locations

Table 450.3(A) - Maximum Rating or Setting of Overcurrent Protection for Transformers Over 600 Volts (as a Percentage of Transformer-Rated Current) (1. - 4.) [4]

5. The impedance of a 3-phase 500kVA transformer rated for 13.8kV-208/120V is unknown although the transformer is located where only qualified employees will provide maintenance. Determine the required overcurrent protection for the transformer.

Calculate transformer's primary and secondary full-load currents

$$I_P = \frac{500kVA\ (500,000VA)}{13,800V \times 1.732} = 20.92A$$

$$I_S = \frac{500kVA\ (500,000VA)}{208V \times 1.732} = 1387.90A$$

Refer to Table 450.3(A) - Supervised Locations

Impedance: Any
Primary Voltage: Over 600 volts

Based on the specifications provided in the question and the requirements listed in Table 450.3(A), the transformer's primary overcurrent device can also provide overcurrent protection for the secondary. In this case, secondary overcurrent protection is not required.

Consider the required overcurrent protection

Primary Protection

If circuit breaker used [300% (3)]   If fuses used [250% (2.5)]
20.92A x 3 = 62.76A. Use a 70A circuit breaker.   20.92A x 2.5 = 52.3A. Use 60A fuses.

Table 450.3(B) - Maximum Rating or Setting of Overcurrent Protection for Transformers 600
                        Volts and Less (as a Percentage of Transformer-Rated Current) (6. - 8.) [3]

6. A 50kVA single-phase transformer rated for 480-240/120V is used for power and lighting circuits in a commercial repair shop. What size overcurrent protection is required to protect the transformer's primary and secondary windings?

Calculate transformer's primary and secondary full-load currents (FLC)

$$I_P = \frac{50kVA\ (50,000VA)}{480V} = 104.2A$$

$$I_S = \frac{50kVA\ (50,000VA)}{240V} = 208.3A$$

Refer to Table 450.3(B)

If method of protection is primary only

Percent increase of primary FLC: [125% (1.25)] (FLC greater than 9 amps)

Consider the required overcurrent protection

Primary Protection

104.2A x 1.25 = 130.25A

Table 450.3(B) **Note 1**, allows the next higher standard rating to be used per NEC 240.6(A), which permits either a 150A nonadjustable circuit breaker or fuses to be used to protect the transformer's primary and secondary windings.

When secondary overcurrent protection is not required, the requirements of **Note 2** does not apply, meaning if secondary overcurrent protection was used, more than 6 circuit breakers or sets of fuses that were grouped in one location could be used. See question No. 9. to reference **Note 2** when it's applicable.

If method of protection is for both primary and secondary protection

Percent increase of primary FLC: [250% (2.5)] (FLC greater than 9 amps)
Percent increase of secondary FLC: [125% (1.25)] (FLC greater than 9 amps)

Consider the required overcurrent protection

Primary Protection

104.2A x 2.50 = 260.5A

Since there are no provisions which allows the results of the calculation to be increased to correspond with a standard rated overcurrent device, a 250A overcurrent device (non-adjustable circuit breaker or fuses) must be used as the transformer's primary overcurrent device.

Secondary Protection

208.3A x 1.25 = 260.38A

Table 450.3(B), **Note 1** allows next standard rated overcurrent device to be used. Use a 300A overcurrent device (non-adjustable circuit breaker or fuses).

7. A single-phase 1kVA transformer is rated for 240V-120V. What size overcurrent protection is required to protect the transformer's primary and secondary windings?

Calculate transformer's primary and secondary full-load currents (FLC)

$$I_P = \frac{1kVA\ (1,000VA)}{240V} = 4.2A$$

$$I_S = \frac{1kVA\ (1,000VA)}{120V} = 8.33A$$

Refer to Table 450.3(B)

If method of protection is primary only

Percent increase of primary FLC: [167% (1.67)] (FLC less than 9 amps)

Consider the required overcurrent protection

Primary Protection

4.2A x 1.67 = 7.01A

Notice, there are no provisions in Table 450.3(B) which allows the use of a next standard rated overcurrent device. Because of such restriction, only fuses can be used [NEC 240.6(A)] because they are available in sizes smaller than 7.01A compared to circuit breakers. As a result, 6A fuses must be used to protect the transformers primary and secondary windings.

If method of protection is for both primary and secondary protection

Percent increase of primary FLC: [250% (2.5)] (FLC less than 9 amps)
Percent increase of secondary FLC: [167% (1.67)] (FLC less than 9 amps)

Consider the required overcurrent protection

Primary Protection

4.2A x 2.50 = 10.5A

Refer to NEC 240.6(A). Again, only fuses can be used because of the next standard rating overcurrent protection restriction. Use 10A fuses to protect the transformer's primary windings.

Secondary Protection

8.33A x 1.67 = 13.91A

The same size fuses (10A) are also required to protect the transformer's secondary windings.

8.   Refer to question No. 28. of Article 430. What size overcurrent protection is required to protect the transformer's primary windings only? Both primary and secondary windings?

Also see question No. 26. of Article 240 (Volume 2).

Per Table 450.3(B) when the primary current of a transformer values 9 amperes or more the primary windings has to be protected at 125% (1.25) of a transformer's primary full-load current.

To determine the required overcurrent devices for the primary windings let's first calculate the transformer's primary full-load current.

$$I_P = \frac{15kVA\ (15{,}000VA)}{460V} = 32.61A$$

Based on the calculated results, the overcurrent protection for the primary windings results to,

$$32.61A \times 1.25 = 40.8A$$

Because the results does not amount to a standard size overcurrent device, the primary overcurrent device is determined based on the provisions of Table 450.3(B), **Note 1** which allows the primary overcurrent device to be increased to the next standard size which is 45A.

To protect both windings

Per Table 450.3(B) when the primary and secondary currents of a transformer values 9 amperes or more the primary windings has to be protected at 250% (2.5) of a transformer's primary full-load current while the secondary windings are required to be protected at 125% (1.25) of a transformer's secondary full-load current.

Having calculated the full-load current of the transformer's primary let's now calculate the transformer's secondary full-load current.

$$I_S = \frac{15kVA\ (15{,}000VA)}{208V} = 72.12A$$

Based on the calculation, the overcurrent protection for the primary windings now results to,

$$32.61A \times 2.5 = 81.53A$$

while the secondary windings results to,

$$72.12A \times 1.25 = 90.15A$$

Because neither result amounts to a standard size overcurrent device, the primary overcurrent device is limited to 80A because there are no provisions to increase it to the next standard size. On the other hand, the secondary overcurrent device can be increased to 100A based on the provisions of Table 450.3(B), **Note 1**.

Just remember, the provisions of Article 450 primary focus is upon the windings of a transformer and its protection. To size and protect conductors connected to a transformer and affiliated devices other NEC sections much be referenced.

The question is now probably asked, "how is it that an overcurrent device can be used to protect the windings of a transformer and not provide the same protection for its secondary conductors or succeeding devices (panelboards, switches, motor controller) where applicable?" The answer is, the limits set for protecting the windings of a transformer are more stringent than the limits set for protecting secondary conductors or succeeding devices. Per Table 450.3(B) the secondary windings are protected based on set percentage values and specific operating conditions, whereas the secondary conductors are allowed to go unprotected based on varying conditions per NEC 240.21(C) and NEC 408.36 which requires panelboards to be protected by an overcurrent device not greater than its rating.

Table 450.3(A) (Transformers over 600 Volts - Any Location), Note 2 - Where Secondary Overcurrent Protection is Required

Just as a set of transformer secondary conductors are permitted to feed separate loads, **Note 2** to Tables 450.3(A) and (B) permits and specify a maximum number of secondary overcurrent devices that one transformer can feed. According to **Note 2**, where secondary overcurrent protection is required, the secondary overcurrent device shall be permitted to consist of not more than six circuit breakers *or* six sets of fuses grouped in one location. Where multiple overcurrent devices are utilized, the total of all device ratings shall not exceed the allowed value of a single overcurrent device. In contrast, only Table 450.3(A), **Note 2** limits the total of all device ratings to that allowed for fuses where both circuit breakers and fuses are used as the secondary overcurrent devices. The six secondary overcurrent devices permitted in **Note 2** take on a similar allowance to that of NEC 230.71(A) where a maximum of six service disconnects are permitted to be grouped in any one location per service.

9. Refer to Figure 240.4(b) of Article 240 (Volume 2). Assume the transformer was rated for 2500kVA at 12.47kV-2300/1328V, three-phase with 2.9 percent impedance. What size overcurrent protection is required to protect the transformer's primary and secondary overcurrent devices, if the secondary devices consist of three individual circuit breakers and three sets of fuses? Consider the transformer being in any location.

Calculate transformer's primary and secondary full-load currents (FLC)

$$I_P = \frac{2500kVA\ (2,500,000VA)}{12.47kV\ (12,470V) \times 1.732} = 115.75A$$

$$I_S = \frac{2500kVA\ (2,500,000VA)}{2300V \times 1.732} = 627.57A$$

Refer to Table 450.3(A)

Location: Any (unsupervised)
Impedance: Not more than 6%
Primary and Secondary Voltage: Over 600 volts

Consider the required overcurrent protection

Primary Protection

If circuit breaker used [600% (6)]
115.75A x 6 = 694.5A. Use a 700A circuit breaker.

If fuses used [300% (3)]
115.75A x 3 = 347.25A. Use 350A fuses.

Secondary Protection

If circuit breaker used [300% (3)]
627.57A x 3 = 1882.71A. Use a 2000A circuit breaker.

If fuses used [250% (2.50)]
627.57A x 2.5 = 1568.93A. Use 1600A fuses.

Table 450.3(A), **Note 2**, allows the secondary feeder conductors from a transformer's secondary to terminate in *not more than* six circuit breakers, six sets of fuses *or* a combination of both type devices not exceeding six where all devices are grouped in one location. Instead of being terminated in one single secondary overcurrent device, the use of multiple sets of overcurrent devices is allowed providing the total value of the sets does not exceed the value of a single overcurrent device.

Based upon the results gathered for one single secondary overcurrent device, the total rating of the six secondary overcurrent devices *cannot exceed* **2000A** if a circuit breaker is used or **1600A** if fuses are used. However, the conclusion of **Note 2** states, if both circuit breakers and fuses are used as the (secondary) overcurrent device(s), the total of the device ratings shall not exceed that allowed for fuses. Therefore, the total rating of the secondary overcurrent devices cannot exceed **1600A**.

Since the question involves six overcurrent devices (3 circuit breakers and 3 fuses), the devices can all be rated the same, that is, 250A, which totals 1500A (250A x 6) leaving an additional 100A to either increase the rating of one of the six devices by 100A or two of the six devices by 50A. In conclusion, any combination of secondary overcurrent device ratings can be used as long as the total does not exceed 1600A.

Again, the only contrast between Table 450.3(A) and (B), **Note 2** is the concluding statement found in Table 450.3(A), **Note 2** other than that, the requirements are the same.

Table 450.3(B) (Transformers 600 Volts and Less), Note 2 - Where Secondary Overcurrent Protection is Required

10. A 480-208/120V, three-phase 45kVA transformer is used to supply two motors and an electric heating furnace. The motors and furnace are all three-phase loads and are protected by

40, 60 and 70 amps fuses respectively in fusible disconnect switches that are grouped in one location. Can each load be protected as stated?

Calculate transformer's primary and secondary full-load currents (FLC)

$$I_P = \frac{45kVA\ (45,000VA)}{480V \times 1.732} = 54.13A$$

$$I_S = \frac{45kVA\ (45,000VA)}{208V \times 1.732} = 124.91A$$

Refer to Table 450.3(B)

Method of protection: primary and secondary

Percent increase of primary FLC: [250% (2.5)] (FLC greater than 9 amps)
Percent increase of secondary FLC: [125% (1.25)] (FLC greater than 9 amps)

Consider the required overcurrent protection

Primary Protection

54.13A x 2.50 = 135.33A

Because the next higher standard overcurrent device is not permitted, a 125A overcurrent device must be used (opposed to a 150A device) to protect the transformer's primary windings.

Secondary Protection

124.91A x 1.25 = 156.14A

Note 1 allows the use of the next higher standard overcurrent device rating per NEC 240.6(A). Therefore, a 175A overcurrent device can be used for the secondary protection.

Table 450.3(B), **Note 2** allows the secondary feeder conductors from a transformer's secondary to terminate in not more than six (6) circuit breakers or six (6) sets of fuses grouped in one location. Instead of being terminated in one single secondary overcurrent device, the use of multiple sets of overcurrent devices is allowed providing the total value of the sets does not exceed the value of a single overcurrent device.

Because the total overcurrent device ratings of the fuses (40A + 60A + 70A = 170A) does not exceed the allowed value of the 175A overcurrent device derived for the secondary protection, the answer to the question is "yes". If the total overcurrent device rating had exceeded the 175A overcurrent device rating, the secondary conductors would then be required to be terminated in a single secondary overcurrent device.

Table 450.3(B), Note 3 - Transformer Equipped with Coordinated Thermal Overload Protection

11. A three-phase, 37.5kVA transformer is equipped with coordinated thermal overload protection and arranged to interrupt the primary current of the transformer. Because it is uncertain whether the transformer's rated impedance is either 5.6 or 6.5, calculate the transformer's primary and secondary overcurrent protection taking both impedance ratings into consideration. The transformer's voltage rating is 480-240/120V.

Calculate transformer's primary and secondary full-load currents (FLC)

$$I_P = \frac{37.5kVA\ (37,500VA)}{480V \times 1.732} = 45.1A$$

$$I_S = \frac{37.5kVA\ (37,500VA)}{240V \times 1.732} = 90.2A$$

Refer to Table 450.3(B)

Method of protection: primary and secondary

Percent increase of primary FLC: [600% (6)] (impedance 6% or less) [**Note 3**]
Percent increase of primary FLC: [400% (4)] (impedance exceeds 6% no more than 10%) [**Note 3**]
Percent increase of secondary FLC: [125% (1.25)] (FLC greater than 9 amps)

Consider the required overcurrent protection

Primary Protection

45.1A x 6 = 270.6A

Note 3 states that the rating of the overcurrent protection cannot exceed 4 times (400%) the rated primary current of the transformer when the transformer has a rated impedance exceeding 6 percent but not greater than 10 percent. A 175A overcurrent device must be used to remain within the 400 percent (400%) restriction.

Primary Protection

45.1A x 4 = 180.4A

Note 3 states that the rating of the overcurrent protection cannot exceed 4 times (400%) the rated primary current of the transformer's full-load current when the transformer has a rated impedance exceeding 6 percent but not greater than 10 percent. A 175A overcurrent device must be used to remain within the 400 percent (400%) restriction.

Secondary Protection

90.2A x 1.25 = 112.75A

Note 1 allows the use of the next higher standard overcurrent device; therefore a 125A overcurrent device can be used for the secondary protection.

450.4(A) - Overcurrent Protection Autotransformers 600 Volts, Nominal, or Less

12. What size overcurrent device is required to protect an autotransformer that has a nameplate input full-load current rating of 37 amps and a voltage rating of 208V - 240V?

NEC 450.4(A) requires the overcurrent device to be sized at no more than 125 percent (1.25) of the transformer's full-load input current.

37 amps x 1.25 = 46.25 amps

Because the input current exceeds 9 amps, the next higher standard rating is permitted. Therefore, a 50A overcurrent device can be used.

ARTICLE 455 – Phase Converters

455.6(A)(1) - Conductors (Ampacity) *and* 455.7(A) - Overcurrent Protection (Variable Loads)

1. A static-phase converter is being used to supply a three-phase motor that serves variable horsepower loads. The input full-load current rating of the phase converter is 65 amps. Determine the size THW copper conductors needed to supply the phase converter and the overcurrent device needed for overcurrent protection.

The ampacity of the supply conductors [NEC 455.6(A)(1)] and the size of the overcurrent device [NEC 455.7(A)] are determined by increasing the phase-converter's nameplate single-phase input full-load amps by 1.25 when the phase-converter supplies variable loads.

$$65 \text{ amps} \times 1.25 = 81.25 \text{ amps}$$

Supply conductors - Use 4 AWG THW copper conductors per Table 310.15(B)(16). Overcurrent device - Use an overcurrent device rated for 90 amps. Next higher standard rated device permitted, NEC 455.7.

455.6(A)(2) - Conductors (Ampacity) *and* 455.7(B) - Overcurrent Protection (Fixed Loads)

2. A single-phase panelboard rated for 240/120V is used to feed a rotary-phase converter which feeds a 480V three-phase panelboard. The three-phase panelboard is used to supply four 3φ motors that are rated for 3, 5, 7.5 and 10 horsepower. In addition to these loads, the three-phase panelboard also supply other 3φ loads totaling 46 amps. The rotary-phase converter is rated for 240V-480V and has a single-phase input current rating of 250 amps. What size THW copper feeder conductors and overcurrent device are needed for the single-phase supply to the rotary-phase converter?

AMPACITY OF FEEDER CONDUCTORS

Because these loads are considered fixed loads, NEC 455.6(A)(2) is referenced to determine the ampacity of the single-phase supply (feeder) conductors. In all, NEC 455.6(A)(2) requires the ampacity of the conductors be no less than 250 percent of the full-load current rating of all motors and other loads served bearing the preceding ampacity requirements are not met. Now let's consider the loads involved.

Total Motor Loads (3φ @ 480V)
FLC per Table 430.250

3HP -	4.8A
5HP -	7.6A
7.5HP -	11.0A
10HP -	14.0A
	37.4A
Other Loads	46.0A
Total	83.4A

After having determined the three-phase loads for this installation there's another factor that has be considered. NEC 455.6(A)(2) concludes by requiring the load current to be increased by the ratio of the output to input voltage of the phase converter when the two voltages differ. Based on this requirement, the load current to be used for determining the ampacity of the feeder conductors is,

$$\frac{480V}{240V} \times 83.4A = 166.8A$$

If the input and output voltages of the phase converter were identical the ampacity rating of the feeder conductors would remain at 83.4A. Now to determine the ampacity of the feeder conductors (at 250 percent of load current),

$$166.8A \times 2.5 = 417A$$

Per Table 310.15(B)(16), 600 kcmil (420A) THW copper conductors are needed as the single-phase feeder conductors.

FEEDER CONDUCTOR'S OVERCURRENT DEVICE

NEC 455.7(B) states that the feeder conductors must be protected by an overcurrent protection device that's in accordance with the ampacity of the conductors when a phase converter supplies specific fixed loads. In this case an overcurrent device rated for 400A is required. However, the section also states that the overcurrent device shall not exceed 125 percent of the phase converter's nameplate single-phase input amperes which is 250A. Considering this requirement,

$$250A \times 1.25 = 312.5A$$

a 300A overcurrent device is required because the overcurrent protection cannot exceed 1.25 of the phase converter's nameplate single-phase input amperes.

A 300A overcurrent device must be used instead of a 400A device.

455.8(C) - Rating (Disconnecting Means)

3. If a disconnect switch was needed for the installation in question No. 1., what size switch would be required?

NEC 455.8 requires the disconnect switch to be no less than 115 percent of the rated maximum single-phase input full-load amperes. As required, the disconnect switch is sized accordingly,

$$65 \text{ amps} \times 1.15 = 74.75 \text{ amps}$$

A standard size switch with a minimum rating of 100 amps is required based on the results of the calculation.

455.8(C)(1) - Current Rated Disconnect (Disconnecting Means) *and* **455.8(C)(2) - Horsepower Rated Disconnect (Disconnecting Means)**

4. A molded-case switch is switch is needed for disconnecting means on the supply side of a rotary phase converter. The phase converter is being used to provide three-phase capabilities for six 208V three-phase motors that operate simultaneous. Three of the motors are Design C motors and rated for 3, 5 and 7½ horsepower while the other three motors share the same operating features other than being Design C motors. What size molded-case switch is required for disconnecting means?

According to NEC 455.8(C), for specific fixed loads the disconnecting means shall be permitted to be selected from either NEC 455.8(C)(1) or (C)(2). Because alternate means are permitted both will be considered.

Per NEC 455.8(C)(1)

The disconnecting means must be a circuit-breaker or molded-case switch with an ampere rating not less than 250 percent (2.5) of the sum of the loads being supplied by the phase converter.

Being that there are no other loads served the sum of the loads will be determined based on the full-load currents of all 3-phase motors.

Full-load current ratings per Table 430.250

$$
\begin{array}{llll}
\text{(2) 3hp -} & 10.6\text{A x 2} = & 21.2\text{A} \\
\text{(2) 5hp -} & 16.7\text{A x 2} = & 33.4\text{A} \\
\text{(2) 7½hp -} & 24.2\text{A x 2} = & \underline{48.4\text{A}} \\
& & 103.0\text{A (total)}
\end{array}
$$

With this data a current rated disconnect switch can be determined.

$$103\text{A x 2.5} = 257.5\text{A}$$

Since the disconnect switch is needed on the supply side (NEC 455.8) of the phase converter a single-phase switch or three-phase switch (excluding the use of one switch pole) with a standard rating exceeding 257.5A is needed.

Per NEC 455.8(C)(2)

The disconnecting means must be a switch with a horsepower rating that has an equivalent locked rotor current that's rated for at least 200 percent (2) of the sum of the loads being supplied by the phase converter.

Being that there are no non-motor loads, the sum of the loads will be determined based on the locked-rotor current of the largest motor per Table 430.251(B) and the full-load currents of all other motors.

Per Table 430.251(B) the Design C, 7½ horsepower motor has the highest locked-rotor current at 140A. The full-load currents of the remaining motors can now be added to this value.

Full-load current ratings per Table 430.250

$$
\begin{aligned}
&(2)\ 3\text{hp} - \quad 10.6\text{A} \times 2 = 21.2\text{A} \\
&(2)\ 5\text{hp} - \quad 16.7\text{A} \times 2 = 33.4\text{A} \\
&(1)\ 7\tfrac{1}{2}\text{hp} - 24.2\text{A} \times 1 = \underline{24.2\text{A}} \\
&\hspace{6.5cm} 78.8\text{A (total)}
\end{aligned}
$$

With this data an equivalent locked-rotor current can be derived to determine the needed horsepower rating of the disconnect switch.

$$(140\text{A} + 78.8\text{A}) \times 2 = 437.6\text{A}$$

Since the disconnect switch is needed on the supply side (NEC 455.8) of the phase converter a single-phase switch is needed. Referring to Table 430.151(A) to size a single-phase-switch based on the calculated locked rotor current, the highest locked-rotor current listed in the table at 208 volts is far less the required 437.6A. Therefore, Table 430.251(B) must be used to size the required horsepower rating of the disconnect switch. Using the first value under the 208-volts listing, which list a locked-rotor current greater than the calculated value requires the horsepower rating of the disconnect switch to be rated for 30 horsepower based on a locked-rotor current of 481A.

As a result, a three-phase disconnect switch rated for 30 horsepower must be used for the disconnecting means which implies that one of the switch poles will not be used.

ARTICLE 460 - Capacitors

460.8(A) - Conductors (1. - 3.) [3]

1. An industrial facility is installing a bank of three-phase capacitors at the service-entrance to improve the facilities power factor rating. The capacitors are rated for 70kVARs at 460 volts (60Hz). What size 75°C copper circuit conductors are needed to supply the capacitor bank?

NEC 460.8(A) requires the ampacity of capacitor circuit conductors to be no less than 135 percent of the capacitor's rated current. Before preceding the capacitor's rated current (I_C) must be determined.

$$I_C = \frac{70kVARs}{460V \times 1.732} = 87.9A$$

The ampacity of the conductors can now be determined.

$$87.9A \times 1.35 = 118.67A$$

As listed in Table 310.15(B)(16), 1 AWG copper conductors which have an ampacity of 130 amps are required to serve as the circuit conductors to supply the capacitor bank.

2. A capacitor rated for 17 amps is connected to the (motor) circuit conductors of a single-phase, 230V, 10hp motor to improve the motor's power factor. The circuit conductors are 6 AWG THWN copper conductors. What size THWN copper conductors are required to connect the capacitor to the circuit conductors?

NEC 460.8(A) requires such conductors to be no less than one-third (1/3) the ampacity of the motor circuit conductors and no less than 135 percent of the rated current of the capacitor.

The 6 AWG motor circuit conductors have an ampacity of 65 amps per Table 310.15(B)(16). One-third (1/3) of 65 amps is 21.67 amps and 135 percent of the capacitor's rated current (17 amps) is 22.95 amps. Considering either requirement, 12 AWG THWN copper conductors which are rated for 25 amps per Table 310.15(B)(16) will suffice to connect the capacitor to the circuit conductors.

3. What size capacitor in units of micro-farads (µf) is needed to produce the rated current in question No. 2.?

The rated current is actually the reactive current produced by the capacitor. With the reactive current and the (assumed) operating voltage of the motor (230V at 60Hz), the capacitance reactance (X_C) can be calculated where,

$$X_C = \frac{230V}{17A} = 13.53\Omega$$

Using the formula,

$$X_C = \frac{1}{2 \times \pi \times f \times C}$$

to derive the formula,

$$C = \frac{1}{X_C \times 2 \times \pi \times f} \text{ (where } \pi = 3.14 \text{ and } f = 60Hz)$$

the value of the capacitor can be calculated.

$$C = \frac{1}{13.53 \times 2 \times 3.14 \times 60Hz}$$

$$= .00019615F \text{ or } 196\mu f \text{ (rounded off)}$$

A capacitor rated for approximately 196µf is needed to produce the capacitor's rated (reactive) current.

460.8(C) - Disconnecting Means

4. What size disconnect switch is required to disconnect the 12 AWG conductors from the capacitor in question No. 2.?

According to NEC 460.8(C)(3), the rating of the disconnecting means must not be less than 135 percent of the rated current of the capacitor (22.95 amps). This being the case, a standard size switch with a minimum rating of 30 amps is required.

ARTICLE 517 - Health Care Facilities

Refer to Worksheet G (Item 7 - MEDICAL EQUIPMENT) for applicable formulas.

517.72(A) - Capacity (Disconnecting Means)

1. Refer to the X-Ray unit in question No. 2. What size disconnecting means is required?

Based on the greater calculated determinant (momentary rating), a disconnecting means with a standard rating exceeding 31.3A is required.

517.73(A)(1) - Branch Circuits (Diagnostic Equipment)

2. The nameplate of an X-Ray unit provides the following data:

Monetary rating (3 seconds)	Long-time rating (12 minutes)
Ampere Rating - 200mA (.2A)	Ampere Rating - 38mA (.038A)
Primary Voltage - 240V (1φ)	Primary Voltage - 240V (1φ)
Secondary Voltage - 75,000V (1φ)	Secondary Voltage - 125,000V (1φ)

What size branch-circuit conductors (75°C-copper) and overcurrent device are required to serve the X-Ray unit? [Use the formula $I = (V_S/V_P) \times A$, where I = load per time rating]

For branch-circuits conductor's and overcurrent devices, NEC 517.73(A)(1) requires the ampacity of branch-circuit conductors and the current rating of overcurrent protective devices to be no less than 50 percent (.50) of the momentary rating or 100 percent (1) of the long-time rating, whichever is greater for such equipment.

Before applying NEC 517.73(A)(1), the first step is to determine the equipment's short and long-time loads in amperes (A).

$$I_{MT} = (75,000V / 240V) \times .200A = 62.5A$$
$$I_{LT} = (125,000V / 240V) \times .038A = 19.8A$$

Momentary Rating @ 50% - 62.5A x .50 = 31.3A
Long-time Rating @ 100% - 19.8A

Because the momentary rating will produce the greater of both loads the equipment's branch-circuit conductors and overcurrent device are determined based on 31.3A. As a minimum, (at 75°C) 10 AWG conductors are required as the branch-circuits conductors which will require a 30A overcurrent device per NEC 240.4(D)(7) and 240.6(A). However, a 30A overcurrent device would cause the equipment to operate less than the required 50 percent of the momentary rating. Therefore, a 35A overcurrent device is needed which would require 8 AWG copper conductors rated for 50A. Although, a 10 AWG copper conductor per Table 310.15(B)(16) has a rated ampacity of 35A, the provisions of NEC 240.4(D)(7) do not allow the use of a 35A overcurrent device.

517.73(A)(2) - Feeders (Diagnostic Equipment)

3. A 208/120V, single-phase feeder supplies a panelboard that serves eight X-Ray units. The momentary rating in amperes for each unit is as follows:

Momentary Amps

X-Ray Unit 1 - 22.7   X-Ray Unit 2 - 38.3   X-Ray Unit 3 - 33.5   X-Ray Unit 4 - 48.2
X-Ray Unit 5 - 18.6   X-Ray Unit 6 - 29.4   X-Ray Unit 7 - 50.8   X-Ray Unit 8 - 28.1

X-Ray units 7 and 8 are used to perform simultaneous biplane examinations. Determine the ampacity of the feeder, the type THW copper conductors needed and the feeder overcurrent protection.

For feeder conductors and overcurrent devices, NEC 517.73(A)(2) requires the ampacity of feeder conductors and the current rating of overcurrent protective devices to be no less than 50 percent (.50) of the momentary demand rating of the largest unit *plus* 25 percent (.25) of the momentary demand rating of the next largest unit *plus* 10 percent (.10) of the momentary demand rating of each additional unit. Where units are used at the same time (simultaneous) to perform biplane examinations, supply conductors and overcurrent devices must be sized at 100 percent (1) of the momentary demand rating of each X-ray unit.

Per NEC 517.73(A)(2)

Simultaneous Use - Units 7 and 8
50.8A + 28.1A = 78.9A

Largest Unit - Unit 4
48.2A x .50 = 24.1A

Next Largest Unit - Unit 2
38.3A x .25 = 9.6A

Others - Units 1, 3, 5 and 6
(22.7A + 33.5A + 18.6A + 29.4A) x .10 = 10.42A

Total
78.9A + 24.1A + 9.6A + 10.42A = 123.02A

The ampacity of the feeder conductors must be no less than 123A, therefore as a minimum, 1 AWG (130A) THW copper conductors per Table 310.15(B)(16) must be used along with a 125A overcurrent device. In accordance with NEC 240.4(B), a 150A overcurrent device could be used.

ARTICLE 520 - Theaters, Audience Areas of Motion Picture and Television Studios, Performance Areas, and Similar Locations

Table 520.44 - Ampacity of Listed Extra-Hard Usage Cords and Cables with Temperature Ratings of 75°C (167°F) and 90°C (194°F) [Based on Ambient Temperature of 30°C (86°F)]

1. A 75°C 5-conductor 10 AWG cable listed for extra-hard usage will be used where 4 of the cable's conductors will be current-carrying. The load being supplied by the cable has a 60 percent (.60) diversity factor. Determine the ampacity of the cable's conductors.

According to the *footnotes* of Table 520.44, when the number of current-carrying conductors in a cable as such exceeds three and the load diversity factor is a minimum of 50 percent (not less than), the ampacity of each conductor shall be reduced.

Referring to Table 520.44, a 10 AWG cable rated for 75°C has an ampacity of 41A. Since four of the cable's conductors are current-carrying the ampacity of the conductors must be reduced by 80 percent (.80) based on the table listed below Table 520.44. Therefore,

$$41A \times .80 = 32.8A$$

The ampacity of the cable's conductors must be limited to 32.8A.

ARTICLE 530 - Motion Picture and Television Studios and Similar Locations

530.18(B) - Feeders (Overcurrent Protection - General)

1. Determine the maximum size overcurrent device that can be used to protect 2/0 AWG THWN copper feeder conductors per NEC 530.18.

Referring to Table 310.15(B)(16), a 2/0 AWG THWN copper conductor has an ampacity of 175 amps. NEC 530.18(B) states that the overcurrent device setting for a feeder shall not exceed 400 percent (4) of the ampacity of the feeder, therefore

$$175A \times 4 = 700A$$

The maximum size overcurrent device for a 2/0 AWG feeder cannot exceed 700 amps. Refer to NEC 240.6(A).

530.19(A) - General (Sizing of Feeder Conductors for Television Studio Sets)

2. What size feeder conductors are required to supply the lighting for a television studio when the total connected lighting load is 185,000VA? The feeder conductors will be fed from a 3φ, 480V source. Use THWN copper conductors rated for 75°C.

NEC 530.19(A) permits the use of the demand factors listed in Table 530.19(A) for sizing permanently installed feeder conductors which will supply studio or stage set lighting. Based on Table 530.19(A):

Connected Lighting Load = 185,000VA

a. 1st 50,000VA @100% = 50,000VA (135,000VA remaining)

b. From 50,001 up to 100,000VA @75%(.75)
(Use 50,000VA of remaining 135,000VA which totals 100,000VA when added to **a.** [1st 50,000VA]).

$$50,000VA \times .75 = 37,500VA$$

c. From 100,001 up to 200,000VA @60%(.60)
(Use 85,000VA which is the remainder when **a.** and **b.** are subtracted from 185,000VA).

$$85,000VA \times .60 = 51,000VA$$

Demand Load (Volt-Amperes [VA])

50,000VA + 37,500VA + 51,000VA = 138,500VA

Demand Load (Amperes [A])

$$\frac{138,500VA}{480V \times 1.732} = 166.59A$$

Based on the calculated demand load and the use of THWN copper conductors, the feeder conductors must be sized no less than 2/0 AWG which has an ampacity of 175 amps.

530.19(B) - Portable Feeders (Sizing of Feeder Conductors for Television Studios Sets)

3. If the total connected load for stage lighting in a motion picture arena is 67,280VA; determine the demand load if a portable feeder is used.

According to NEC 530.19(B), a demand factor of 50% (.50) of the maximum possible connected load shall be permitted (to determine the demand load) for all portable feeders.

Demand Load (Volt-Amperes [VA]) of Portable Feeder

$$67,280VA \times .50 = 33,640VA$$

ARTICLE 550 - Mobile Homes, Manufactured Homes, and Mobile Home Parks

550.12(A) - Lighting (Branch Circuits)

1. Refer to question No. 2. How many 15A lighting circuits are required for this mobile home?

Using the formula given in NEC 550.12(A), where the dimensions of the mobile home are 30' x 20', the number of 15A lighting branch circuits amounts to:

$$\frac{3 \times 30' \times 20'}{120V \times 15A} = 1$$

As a minimum, one 15A lighting branch circuit is required.

550.18 - Calculations (2. - 3.) [2]

2. The outside measurements of a mobile home is 30' x 20'. The mobile home's electrical load consist of the following:

> Two (2) small appliance circuits
> Laundry circuit
> 3800W electric furnace - 240V
> 1.5 Ton A/C unit @ 7.6A - 240V
> 1/6 HP blower motor @ 1.6A (NPC) - 240V
> 360VA dishwasher - 120V
> 3.5kW electric range - 240/120V

The mobile home will be fed from a 240/120V source. What size power supply is needed to supply the mobile home's electrical load?

Based on NEC 550.18(A)**(1) - (5)**

(1) Lighting VA
    30ft x 20ft x 3VA/ft^2 = 1800VA

(2) Small Appliance VA [2 minimum - NEC 550.12(B)]
    1500VA x 2 circuits = 3000VA

(3) Laundry Circuit VA [1 minimum - NEC 550.12(C)]
    1500VA x 1 circuit = 1500VA

(4) Total VA (1) - (3)
    1800VA + 3000VA + 1500VA = 6300VA

(5) Net VA

1st 3000VA @ 100%	= 3000VA
Remainder (3300VA)* @ 35%(.35) – 3300VA x .35	= 1155VA
*(6300VA – 3000VA = 3300VA)	4155VA

Amperes per leg - 4155VA/240V = 17.31A

Based on NEC 550.18(B)**(1) - (5)**

		Amperes per Leg		
		L₁	**L₂**	**N**
(1)	Lighting, Small Appliance, Laundry [NEC 550.18(A)(5)]	17.31A	17.31A	17.31A
(2)	Nameplate Amperes for Motors and Heater Loads			
	Motors			
	1.6A Blower (240V)	1.60A	1.60A	--
	Heater/AC Load			
	3800W Electric Furnace (Larger than AC load) 3800W / 240V = 15.83A	15.83A	15.83A	--
(3)	Largest Motor Blower - 1.6A x .25 = .4A	.40A	.40A	--
(4)	Appliances 360VA Dishwasher 360VA/120V = 3A	--	3.00A	3.00A
(5)	Freestanding Range 3500W/240V x .8 (Table) = 11.67A 11.67A x .70 = 8.17A [Neutral NEC 220.61(B)(1)]	11.67A	11.67A	8.17A
	Total	46.81A	49.81A	28.48A

Considering the largest ampere rating per leg (49.81A), a 50A mobile home power supply cord or a permanently installed feeder can be used to supply the mobile home based on NEC 550.10(A).

3. A 2365SF mobile home unit is being supplied by a 240/120V, 1φ service. What size service and neutral conductors are required to supply the unit, if the following loads are connected?

120V	240V
Small appliance circuits (3)	Electric Furnace - 18kW
Laundry circuit	A/C unit - 8400VA
Dishwasher - 700VA	Blower - 11.83A (NPC)
Garbage disposal - 420VA	Water Heater - 4.5kW
Heater-Vent-Light - 1300W	Dryer - 5kW
	Range - 8kW
	Outdoor spa - 2.5kW (future)

How many lighting circuits are required for this mobile home?

Based on NEC 550.18(A)**(1) - (5)**

(1) Lighting VA
$$2365ft^2 \times 3VA/ft^2 = 7095VA$$

(2) Small Appliance VA [2 minimum - NEC 550.12(B)]
$$1500VA \times 3 \text{ circuits} = 4500VA$$

(3) Laundry Circuit VA [1 minimum - NEC 550.12(C)]
$$1500VA \times 1 \text{ circuit} = 1500VA$$

(4) Total VA (1) - (3)
$$7095VA + 4500VA + 1500VA = 13,095VA$$

(5) Net VA

1st 3000VA @ 100%	= 3000.00VA
Remainder (10,095VA)* @ 35%(.35) – 10,095VA x .35	= 3533.25VA
*(13,095VA – 3000VA = 10,095VA)	6533.25VA

Amperes per leg - 6533.25VA/240V = 27.22A

Based on NEC 550.18(B)(1) - (6)

		Amperes per Leg		
		L₁	**L₂**	**N**
(1)	Lighting, Small Appliance, Laundry VA [NEC 550.18(A)(5)]	27.22A	27.22A	27.22A
(2)	Nameplate Amperes for Motors and Heater Loads			
	Motors			
	Blower (240V)	11.83A	11.83A	--
	Heater/AC Load			
	18kW Electric Furnace (Larger than AC load) 18,000W / 240V = 75A	75.00A	75.00A	--
(3)	Largest Motor Blower -11.83A x .25 = 2.96A	2.96A	2.96A	--
(4)	Appliances (More than three) Dishwasher 700VA/120V x .75 = 4.38A	4.38A	--	4.38A
	Garbage Disposal 420VA/120V x .75 = 2.63A	--	2.63A	2.63A
	Heat-Vent-Light 1300W/120V x .75 = 8.13A	8.13A	--	8.13A
	Water Heater 4500W/240V x .75 = 14.06A	14.06A	14.06A	--
	Dryer 5000W/240V x .75 = 15.63A 15.63A x .70 = 10.94A [Neutral NEC 220.61(B)(1)]	15.63A	15.63A	10.94A
(5)	Freestanding Range 8000W/240V x .8 (Table) = 26.67A 26.67A x .70 = 18.67A [Neutral NEC 220.61(B)(1)]	26.67A	26.67A	18.67A
(6)	Anticipated Load Outdoor Spa 2500W/240V = 10.42A	10.42A	10.42A	--------
	Total	196.30A	186.42A	71.97A

Considering the largest ampere rating per leg (196.30A) a 200A service would be required. Referring to NEC 310.15(B)(7) and Table either 2/0 AWG copper or 4/0 AWG aluminum conductors can be used as the service-entrance conductors.

Referring to Table 310.15(B)(16) (75°C), a 4 AWG copper or a 3 AWG aluminum conductor can be used as the service neutral.

However, as a minimum, a 2 AWG aluminum conductor must be used as the service neutral conductor instead of a 3 AWG aluminum conductor to be in compliance with NEC 250.24(C)(1). Refer to Table 250.66.

As for the number of either 15A or 20A lighting circuits,

15A - $\dfrac{7095VA}{120V \times 15A}$ = 3.94 rounded up to **4**   **20A** - $\dfrac{7095VA}{120V \times 20A}$ = 2.96 rounded up to **3**

As a minimum, either four 15A or three 20A lighting branch circuits are required.

550.18(C) - Optional Method of Calculation for Lighting and Appliance Load (Calculations) (4. - 5.) [2]

4. Use the optional load calculation to determine the power supply needed to supply the mobile home's electrical load in question No. 2.

NEC 550.18(C) states that the optional method for calculating lighting and appliance loads per NEC 220.82 shall be permitted.

1. General Lighting and Receptacle Loads <NEC 220.82(B)(1)> - (Open porches, garages, unused or unfinished spaces not adaptable for future use not included)

$$30ft \times 20ft \times 3VA/ft^2 = 1800VA$$

2. Small Appliance (Portable) **and Laundry Circuit Load** <NEC 220.82(B)(2)>

$$1500VA \times 3 = 4500VA$$

3. Appliances and Motor Loads <NEC 220.82(B)(3) and (4)>

Appliances and Motors Loads	VA rating
Blower (240V x 1.6A)	384VA
Dishwasher	360VA
Range	3500VA
Total Appliances VA rating =	4244VA

4. Apply Demand Factor <NEC 220.82(B)>

$$\text{TOTAL} = \underline{10,544\text{VA}}$$
[Add loads (1) – (3)]

a. First 10kVA (10,000VA) or less of above TOTAL = 10,000.0VA

b. 544VA (Remainder of TOTAL) x .40   =   217.6VA

c. Total   = 10,217.6VA

5. Heating And Air-Conditioning Load <NEC 220.82(C)>

a. AC - 240V x 7.6A = 1824VA
b. and c. - NA
d. Heat - 3800W x .65 = 2470VA
e. and f. - NA

TOTAL DEMAND LOAD
(Add lines **4c** and **5** [the largest selection])

LINE LOAD

4c. General Lighting and Receptacle, Small
Appliances and Laundry Circuit Loads
and Appliances   =   10,217.6VA

5. Heating and Air-Conditioning (AC) Equipment   =   2470.0VA

TOTAL DEMAND LOAD =   12,687.6VA

DWELLING'S OPERATING LINE VOLTAGE - 240V
(Given operating voltage or as determined per test examination)

CALCULATE MINIMUM LINE AND NEUTRAL LOADS
(Divide Total Demand Load [VA] by operating line voltage [V])

LINE LOAD = 12,687.6VA / 240V = **52.87A**

SIZING FEEDER/SERVICE CONDUCTORS <NEC 220.82(A)>

The optional method yield results that are slightly higher than the standard calculation used in NEC 550.18. Per NEC 220.82(A) the ampacity of the feeder or service conductors supplying this mobile home must be no less than 100 amps.

Based on the results of the optional method, a 50A power supply cord could not be used. NEC 550.10(I) provides guidelines to follow when the calculated load exceeds 50 amps. Per NEC

230.79(C) the service disconnecting means shall have a rating not less than 100 amperes, 3-wire for one-family dwelling.

5. Use the optional load calculation to determine the size service and neutral conductors needed to supply the mobile homes electrical load in question No. 3.

NEC 550.18(C) states that the optional method for calculating lighting and appliance loads shown in NEC220.82 shall be permitted.

1. **General Lighting and Receptacle Loads** <NEC 220.82(B)(1)> - (Open porches, garages, unused or unfinished spaces not adaptable for future use not included)

$$2365ft^2 \times 3VA/ft^2 = 7095VA$$

2. **Small Appliance** (Portable) **and Laundry Circuit Load** <NEC 220.82(B)(2)>

$$1500VA \times 4 = 6000VA$$

3. **Appliances and Motor Loads** <NEC 220.82(B)(3) and (4)>

Appliances and Motors Loads	VA rating
Blower (240V x 11.83A)	2839.2VA
Dishwasher	700.0VA
Disposal	420.0VA
Dryer	5000.0VA
Range	8000.0VA
Water Heater	4500.0VA
Heater-Vent-Light	1300.0VA
Outdoor Spa	2500.0VA
Total Appliances VA rating =	25,259.2VA

4. **Apply Demand Factor** <NEC 220.82(B)>

$$TOTAL = \underline{38,354.2VA}$$
[Add loads (1) – (3)]

a. First 10kVA (10,000VA) or less of above TOTAL = 10,000.00VA

b. 28,354.2VA (Remainder of TOTAL)   x .40   = 11,341.68VA

c. Total   = 21,341.68VA

5. **Heating And Air-Conditioning Load** <NEC 220.82(C)>
   a. AC = 8400VA
   b. and c. - NA

d. Heat - 18,000W x .65 = 11,700VA

e. and f. - NA

TOTAL DEMAND LOAD
(Add lines **4c** and **5** [the largest selection])

LINE LOAD

4c. General Lighting and Receptacle, Small
Appliances and Laundry Circuit Loads
and Appliances   21,341.68VA

5. Heating and Air-Conditioning (AC) Equipment   11,700.00VA

TOTAL DEMAND LOAD = 33,041.68VA

DWELLING'S OPERATING LINE VOLTAGE - 240V
(Given operating voltage or as determined per test examination)

CALCULATE MINIMUM LINE AND NEUTRAL LOADS
(Divide Total Demand Load [VA] by operating line voltage [V])

LINE LOAD = 33,041.68VA / 240V = **137.67A**

SIZING FEEDER/SERVICE CONDUCTORS <NEC 220.82(A), 310.15(B)(7) and Table 310.15(B)(7)> Ampacity of Service or Feeder conductors must be 100A or greater.

Based on the calculated line load (137.67A) a 150A service would be required as a result of the optional method opposed to a 200A service per standard calculation. Referring to NEC 310.15(B)(7) and Table either 1 AWG copper or 2/0 AWG aluminum conductors can be used as the service-entrance conductors.

SIZING NEUTRAL CONDUCTOR <NEC 220.61, 220.82(A), 230.42(C), 250.24(C)(1), 310.15(B)(5) and 310.15(B)(7)>

The neutral conductor is sized based on the calculated neutral load in question No. 3.

550.31 - Allowable Demand Factors (III. Services and Feeders) (6. - 7.) [2]

6. A mobile home park is being constructed to accommodate 50 mobile homes. The electrical load for the largest mobile home is estimated to be approximately 13,500VA. Determine the service demand load for the mobile home park if supplied by a 240/120V single phase source.

NEC 550.31 provides two criteria's for calculating the service load for a mobile home park. As a minimum, 16,000VA must be used for determining the service load for each mobile home lot. This means, when the electrical load for a mobile home is less than 16,000VA the mobile home's electrical load must still be based on the 16,000VA, regardless.

Where provisions of NEC 550.18 are used to calculate the actual electrical load for the largest mobile home that can be placed on a lot, the actual calculated load must be used if it exceeds 16,000VA.

Once the electrical load is determined per criteria, the demand factors listed in Table 550.31 is permitted in order to derive the mobile home park's service demand load.

$$\text{Estimated electrical load} = 13,500\text{VA}$$
$$\text{Minimum electrical load} = 16,000\text{VA (larger)}$$

Demand Factor per Table 550.31 = 23%(.23) based on 50 mobile homes

$$16,000\text{VA} \times 50 \times .23 = 184,000\text{VA}$$

The service demand load for the mobile home park is 184,000VA.

7. Determine the service demand for a mobile home park based on 20 lots if the mobile home given in question No. 3. is the typical size mobile home that will be placed on a lot.

Using the largest calculated ampere rating per leg based on 240V, the mobile homes volt-amperes rating is,

$$240\text{V} \times 196.30\text{A} = 47,112\text{VA (calculated load)}$$

Use larger, calculated load exceeds 16,000VA.

$$47,112\text{VA} \times 20 \times .25 = 235,560\text{VA}$$

The service demand load for the 20 lot mobile home park based on the 2365SF mobile home as a typical size is 235,560VA.

ARTICLE 551 - Recreational Vehicles and Recreational Vehicle Parks

551.20(B) - Voltage Converter (120-Volt Alternating Current to Low-Voltage Direct Current)

1. Determine the voltage converter rating for a recreational vehicle that has a total connected load of 95A.

NEC 551.20(B) provides a formula for determining the rating of a voltage converter.

The 1st 20 amperes of the load at 100 percent	= 20.00A
The 2nd 20 amperes of the load at 50 percent(.50) (20A x .5)	= 10.00A
All loads above 40 amperes at 25 percent(.25)	
(Remaining load 55A) (55A x .25)	= 13.75A
	43.75A

The rating of the voltage converter has to be rated for at least 43.75 amps.

551.73(A) - Basis of Calculations (Calculated Load)

2. The electrical service (240/120V-1φ) for a recreational vehicle (RV) park was recently upgraded to provide more space for RV tenants. The park now has the electrical capacity to supply the following sites:

8 - 20A supply facilities for tents
15 - 20A supply facilities
27 - 20A and 30A supply facilities
36 - 50A supply facilities

In addition, the following house loads were added to the electrical service:

20' x 20' office	6.3A @ 240V - 1 Ton AC unit
10 - duplex receptacles	240V - 5kW furnace
2.8A @ 120V water fountain	120V - 1800VA electric sign
240V - 7.5A copier	240V - 3.8kW water heater

Swimming Pool	Security Lighting
2 - 240V, 1φ, 5HP pumps	5 - 175W Metal Halide fixtures - 120V @ 1.22A
	30 - 250W Metal Halide fixtures - 120V @ 1.83A

Determine the service demand for the recreational vehicle park and house loads.

SITE FACILITIES DEMAND LOAD

The volt-amperes (VA) listed in NEC 551.73(A) must be used for the given site to determine the site facilities total demand load.

20A supply facilities for tents -	8 x 600VA	=	4,800VA
20A supply facilities -	15 x 2400VA	=	36,000VA
20A and 30A supply facilities -	27 x 3600VA	=	97,200VA
50A supply facilities -	36 x 9600VA	=	345,600VA
	86		483,600VA

Applying Table 551.73(A)

Number of RV sites = 86   Demand Factor = 41%(.41)
483,600VA x .41 = 198,276VA

HOUSE LOAD

The following loads must be sized separately and added to the RV park's demand load per NEC 551.73(D). House loads calculated per Article 220, Part III.

Office
General Lighting Load [Table 220.12] - 20' x 20' x 3.5VA x 1.25*	=	1750VA
Receptacles [NEC 220.14(I)] - 10 x 180VA	=	1800VA

AC = 240V x 6.3A = 1512VA (omit) (NEC 220.60)

Furnace - (5kW)	=	5000VA
Water Fountain - 120V x 2.8A	=	336VA
Copier - 240V x 7.5A	=	1800VA
Electric Sign [NEC 220.14(F)] -1800VA x 1.25*	=	2250VA
Water Heater - 3.8kW	=	3800VA
		16,736VA

Swimming Pool
Pumps - (Table 430.248-5HP Motor @ 240V-FLC = 28A)
240V x 28A x 2   = 13,440VA

Security Lights
Metal Halide Fixtures (NEC 220.18(B)

120V x 1.22A x 5 x 1.25*	=	915VA
120V x 1.83A x 30 x 1.25*	=	8235VA

*NEC 230.42(A)

Largest Motor
5HP Pump - 240V x 28A x .25   = 1680VA
(HOUSE LOAD) 41,006VA

TOTAL LOAD (SITE FACILITIES and HOUSE LOAD)

198,276VA + 41,006A = 239,282VA

The service demand for the recreational vehicle park and house load is 997A (239,282VA/240V) which requires a 1000A service as a minimum. If required to size the service conductors the ampacity of the conductors must be either equal to or exceed the rating of the overcurrent device (if the single rating is 1000A) according to NEC 240.4(C). Also, NEC 551.73(D) requires the feeder's neutral conductors to have an ampacity not less than the ungrounded feeder conductors.

ARTICLE 552 - Park Trailers

552.10(E)(1) - Low-Voltage Systems (Overcurrent Protection)

Meet similar requirements as NEC 240.4(D) (Volume 2).

552.46(B)(1) - Lighting (Branch Circuits)

Same application as 550.12(A), see question No. 1.

ARTICLE 555 - Marinas and Boatyards

555.12 - Load Calculations for Service and Feeder Conductors (1. - 2.) [2]

1. A marina consisting of 40 boat slips (land pier for boats) is equipped with the following outlets:

8 - 30A/120V receptacles

8 - 30A/240-120V receptacles

8 - 50A/240-120V receptacles

8 - 60A/240-120V receptacles

8 - 20A/120V and 40A/240-120V receptacles

Based on the number of receptacles, what size 240/120V, single-phase service is required to supply the marina?

Let's begin calculating the 240/120V receptacle loads followed by the 120V receptacle loads.

240/120V receptacles (for dual voltage line-to-line/neutral loads)

30A receptacles - 30A x 240V x 8 =   57,600VA (57.6kW)
[+]40A receptacles - 40A x 240V x 8 =   76,800VA (76.8kW)
50A receptacles - 50A x 240V x 8 =   96,000VA (96.0kW)
60A receptacles - 60A x 240V x 8 = 115,200VA (115.2kW)

345,600VA

[+]Table 555.12 **Note 1**, states where shore power accommodations provide two receptacles specifically for an individual boat slip and these receptacles have different voltages, only the receptacle with the larger kilowatt demand shall be required to be calculated.

120V receptacles (for single voltage line-to-neutral loads)

30A receptacles - 30A x 120V x 8 = 28,800VA (28.8kW)

Because the eight 30A receptacles will serve line-to-neutral connected loads they can be equally balanced across L_1 and L_2 of the 240/120V single-phase service.

TOTAL CONNECTED LOAD

345,600VA + 28,800VA = 374,400VA

Table 555.12 permits the total connected load to be reduced per number of receptacles. Although there are a total of 48 receptacles, only 40 receptacles are considered (the 8-20A/120V receptacles not considered per **Note 1**). Based on 40 receptacles, the total connected load can be reduced by 60 percent (.60) to achieve the total demand.

Therefore,

$$\frac{374,400VA}{240V} = 1560A$$

1560A x .60 = 936A (Total Demand Amperes)

The total demand amperes (936A) reflects the need for a 1000A service.

2. Determine the final demand amperes, if each of the 40 boat slips in question No. 1. were equipped with an individual kilowatt-hour meter.

Table 555.12 **Note 2**, states where individual kilowatt-hour meters are installed for each boat slip of a facility; the total demand amperes may be multiplied by 90 percent (.90) to achieve the final demand amperes.

As a result, the final demand amperes in this case would be 842.4A (936A x .90).

ARTICLE 610 - Cranes and Hoists

Refer to Worksheet G (Item 6C - MOTOR LOADS) (Volume 4) for applicable formulas pertaining to **Article 610**.

Table 610.14(A) - Ampacities of Insulated Copper Conductors Used with Short-Time Rated Crane and Hoist Motors. Based on Ambient Temperature of 30°C (86°F). Up to Four Conductors in Raceway or Cable. Up to three ac or four dc Conductors in Raceway or Cable (1. - 2.) [2]

1. Six 10 AWG SIS conductors installed in a raceway are energized simultaneously to energize a hoist that's operated by a *dc* motor. Applying a 30-minute time rating, determine the allowable ampacity of the conductors.

Referring to the heading (above 90°C column) of Table 610.14(A) and (footnote) **1**, when 5 - 8 power conductors (in this case *dc*) are installed in a raceway and energized simultaneously, the ampacity of each power conductor must be reduced to 80 percent (.80) of the value shown in the table.

Having a 30-minute time rating, Table 610.14(A) list the ampacity of a 10 AWG SIS conductor at 52 amps. The conductor's new allowable ampacity when reduced is 41.6 amps.

$$52 \text{ amps x } .80 = 41.6 \text{ amps}$$

2. Five 3 AWG FEP conductors rated for 125°C are installed in 1½" EMT. The conductors will be used to operate a 240V, 3φ crane motor that has a 15-minute time rating. Determine the allowable ampacity of the conductors.

Referring to the heading (above 125°C column) of Table 610.14(A) and (footnote) **2**, when 4-6 power conductors with a 125°C temperature rating are installed in a raceway and energized simultaneously the ampacity of each power conductor must be reduced to 80 percent (.80) of the value shown in the table.

Since the conductors will be used to energize an *ac* motor that has a 15-minute time rating, the ampacity (183A) listed in the 125°C column at 30-minutes must be increased by 12 percent (.12) according to the **Note** (footnote) and then reduced by 80 percent (.80) to derive the conductor's allowable ampacity. Therefore,

$$183A \times 1.12 \times .80 = 163.97A$$

610.14(B) - Secondary Resistor Conductors (Rating and Size of Conductors)

3. A 480V, 3ϕ wound-rotor motor is used to operate a crane. The motor's nameplate displays a 30-minute duty cycle and a secondary full-load current of 78A. A controller which controls the speed of the motor is connected to a resistor bank through a set of secondary resistor conductors. Determine the size of the secondary resistor conductors at 75°C, where a 15/30 seconds on/off duty cycle is used.

To determine the ampacity of the secondary conductors, NEC and Table 610.14(B) must be referenced to determine the appropriate percent factor per on/off duty cycle. Based on the given duty cycle (15/30s), the motor's secondary full-load current must be multiplied by 75 percent (.75) to determine the ampacity of the conductors.

$$78A \times .75 = 58.5A$$

Table 610.14(A) can now be referenced according to NEC 610.14(B) to size the secondary resistor conductors based on an ampacity of 58.5A.

Referring to the 75°C column of Table 610.14(A) based on a 30-minute rating, 8 AWG conductors which have an ampacity of 60 amps can be used (as a minimum) for the secondary resistor conductors.

Note the conductor ampacities listed in Table 610.14(A) only reference copper conductors.

For an additional question pertaining to this subject matter, refer to question No. 23. of Article 430.

610.14(E)(1) - Single Motor (Calculation of Motor Load)

4. The nameplate of a 230V, 3ϕ, 40HP, squirrel cage motor list the motor's full-load current at 95 amps. The motor is used for a hoist in a machine shop for lifting various size loads. What size 75°C conductors are required to supply the motor if the motor has a 15-minute duty rating?

According to NEC 610.14(E)(1), where one motor is used on a crane or hoist, the ampacity of the motor's supply (power) conductors must be based on 100 percent (1) of the motor's nameplate full-load amperes opposed to the motor's full-load current per NEC 430. Therefore, according to NEC and Table 610.14(A), a set of 5 AWG conductors which have an ampacity of 95 amps based on 75°C can be used to supply the hoist motor.

However, because this motor has a 15-minute duty rating the **Note** to Table 610.14(A) requires the ampacity of a conductor for a motor with a 30-minute duty rating to be increased by 12 percent (.12) to determine the ampacity of a conductor for a motor that has a 15-minute duty rating.

As a result, a set of 6 AWG conductors that's rated for 75°C with a 30-minute allowable ampacity of 86 amps can be used instead.

Increasing this ampacity by 12 percent (.12) yields a conductor ampacity of 96.32A (86A x 1.12) which exceeds the motor's nameplate full-load amperes at 100 percent.

610.14(E)(2) - Multiple Motors on Single Crane or Hoist (Calculation of Motor Load)

5. A crane consisting of three 460V, 3ϕ motors is used in a steel mill for moving raw and finished products. The motor's nameplate lists the following information:

Motor	HP	Full-load Amperes (FLA)	Time-Rating
1	40	45	60 min
2	25	28	30 min
3	25	28	30 min

Note: The two 25HP motors are used simultaneously

What size RHW feeder conductors are required to supply the group of motors? What size RHW conductors are required to supply each individual motor?

SUPPLY CONDUCTORS

According to NEC 610.14(E)(2), to determine the minimum ampacity of the power supply conductors on a single crane, the nameplate full-load ampere rating of the largest motor *or* group of motors for any single crane motion, *plus* 50 percent (.50) of the nameplate full-load ampere rating of the next largest motor *or* group of motors must be applied.

The largest motor = 40HP @ 45A
Group of motors for any single crane motion = (2) 25HP @ 56A (combined)

Because the two 25HP motors will be used simultaneous, thus producing one single crane motion [see last sentence of NEC 610.42(A)], their total combined nameplate full-load ampere ratings will exceed the nameplate full-load ampere rating of the 40HP motor which is the largest single motor. Therefore, 100 percent (1) of the combined nameplate full-load ampere rating of the group of 25HP motors must be added to 50 percent (.50) of the 40HP nameplate full-load ampere rating which is technically the next largest motor.

56A + (45A x .50) = 78.5A

NEC 610.14(E)(2) concludes by requiring the ampacity of the power supply conductors per Table 610.14(A) to be based on the motor with the longest time rating, which is the 40HP motor. Referring to the 75°C column of Table 610.14(A) which list conductors with RHW insulation, the ampacity of the power supply conductor is based on a 60-minute allowable ampacity which requires 5 AWG conductors (85A) as the power supply conductors per calculated ampacity (78.5A).

INDIVIDUAL CONDUCTORS

Again, NEC 610.14(E)(1) is used to size conductors per single motor at 100 percent of each motor's nameplate full-load ampere rating per motor's time rating. Observe,

Motor	FLA	Time-Rating	Individual (Power) Conductors
40HP	45A	60 minutes	Use 8 AWG RHW rated for 55A
25HP (2)	28A	30 minutes	Use 12 AWG RHW rated for 33A

610.14(E)(3) - Multiple Cranes or Hoists On a Common …. (Calculation of Motor Load)

6. The following crane and hoist loads will be supplied by a common conductor system. All cranes and hoist motors are 3ϕ, 208V.

CRANE 1

Motor	HP	FLA (Nameplate)	Time-Rating
1	10	26A	30 minutes
2	15	38A	30 minutes

CRANE 2

Motor	HP	FLA (Nameplate)	Time-Rating
1*	7.5	19A	15 minutes
2*	7.5	19A	15 minutes
3	20	51A	30 minutes

HOIST 1

Motor	HP	FLA (Nameplate)	Time-Rating
1	5	11A	15 minutes

HOIST 2

Motor	HP	FLA (Nameplate)	Time-Rating
1*	2	5.7A	60 minutes
2*	3	8.3A	60 minutes

*simultaneous use

What size common conductors are required to supply the crane and hoist loads?

Where multiple cranes, hoists or both, are supplied by a common conductor system, NEC 610.14(E)(3) requires the minimum ampacity per crane or hoist motor(s) to be determined based on NEC 610.14(E)(1) for a single crane or hoist motor and NEC 610.14(E)(2) for multiple motors on a single crane or hoist. Once the ampacities are determined they are added together and the sum is multiplied by the appropriate demand factor in Table 610.14(E).

Per NEC 610.14(E)(1) - Single Motor

HOIST 1	Minimum Motor Ampacity (per HOIST 1)
Motor 1- 5HP (1)	100 percent of nameplate full-load amperes (NPFLA) = 11A

Per NEC 610.14(E)(2) - Multiple Motors on Single Crane or Hoist

CRANE 1
Motor 2 - Largest Motor   = 38A (2)
Motor 1 - Next largest motor   = 26A (3)

$$38A + (26A \times .50) = 51A$$

Minimum Motor Ampacity (per CRANE 1) = 51A

CRANE 2
Motors 1 and 2 - Combined NPFLA (simultaneous)* = 38A (4)
Motor 3 - NPFLA   = 51A (5)

  Motor 3 - Largest Motor   = 51A
  Motors 1 and 2 - Group of Motors   = 38A

$$51A + (38A \times .50) = 70A$$

Minimum Motor Ampacity (per CRANE 2) = 70A

HOIST 2
Motors 1 and 2 - Combined NPFLA (simultaneous)** = 14A (6)

Minimum Motor Ampacity (per HOIST 2) = 14A

   (1) TOTAL MINIMUM MOTOR AMPACITY
        11A (HOIST 1) + 51A (CRANE 1) + 70A (CRANE 2) + 14A (HOIST 2) = 146A

   (2) Per Table 610.14(E)
        Number of Cranes/Hoist (Motors) = 6   Demand Factor = 81 percent (.81)
   (3) Size of Common Conductor
        146A A x .81 = 118.26A

**Combination motor loads per crane [2] and hoist [2] treated as single motors.

7. Assuming the motors in question No. 5. are all squirrel cage motors, what size branch circuit and feeder overcurrent devices are needed to protect the motors if inverse time circuit breakers (Table 430.52) are used?

BRANCH CIRCUIT OVERCURRENT DEVICES

NEC 610.42(A) states where two or more motors operate a single motion, the sum of the nameplate current ratings must be considered as that of a single motor.

Because the two 25HP motors will operate simultaneously and produce a single motion between the two motors, the two nameplate currents* are added together and treated as one motor.

$$56A (28A + 28A) \times 1.75 = 98A$$

Using the next standard size circuit breaker allowed [per NEC 430.52(C)(1), *Exception 1* and 240.6(A)]** a **100A** inverse time circuit breaker can be used.

The size inverse time circuit breaker used for the 40HP motor is as such,

$$45A \times 1.75 = 78.75A$$

Using the next standard size circuit breaker** an **80A** inverse time circuit breaker can be used.

*Although NEC 610.42(A) reference the use of Table 430.52 while Table 430.52 reference the use of a motor's *full-load current*, as it pertains to Table 430.52 in this situation, only the listed percentage values are applied and not the motor's *full-load current* per values listed in the full-load current tables of Article 430. In conclusion, when selecting overcurrent protective devices per NEC 610.42(A), nameplate current ratings are applied along with the percentage values of Table 430.52.

FEEDER CIRCUIT OVERCURRENT DEVICE

NEC 610.41(A) states that the feeder overcurrent device must not be greater than the largest rating or setting of any branch circuit protective device *plus* the sum of the nameplate ratings of all other loads. The use of the demand factors of Table 610.14(E) does not apply because question No. 5. only involves one crane.

The largest branch overcurrent device in this case is **100A**. Therefore,

$$100A + 45A = 145A$$

Remember, NEC 610.41(A) states that the feeder overcurrent device cannot be greater than the largest rating or setting of any branch-circuit protective device *plus* the sum of the nameplate ratings of all other loads (45A) with application of the demand factors from Table 610.14(E).

— The 2011 National Electrical Code Book of In-Depth Calculations — Volume 3 —

Although there are a total of three motors, the two 25HP motors are treated as one leaving only the 40HP motor which again combines to represent one crane. As a result, the demand factors of Table 610.14(E) are not applicable. Being the case, a 125A inverse-time circuit breaker must be used because the next standard size circuit breaker is not allowed.

8. Assume the branch-circuit overcurrent device ratings for the following crane/hoist motor(s).

CRANE 1

Motor	Overcurrent Device Rating	NPFLA
1 (1)	50A	26A
2 (2)	70A	38A

CRANE 2

Motor	Overcurrent Device Rating	NPFLA
1*	70A	19A
2* (3)	--	19A
3 (4)	90A	51A

HOIST 1

Motor	Overcurrent Device Rating	NPFLA
1*	25A	5.7A
2* (5)	--	8.3A

HOIST 2

Motor	Overcurrent Device Rating	NPFLA
1 (6)	20A	11A

*simultaneous use

What size overcurrent device is required to protect the common conductor system supplying the crane hoist loads?

In accordance with NEC 610.41(A), Motor 3 of Crane 2 has the largest overcurrent device, 90A.

NEC 610.41(A) now requires the sum of the nameplate ratings of all other loads to be totaled and multiplied by the applicable demand factor per Table 610.14(E).

The sum of the nameplate ratings of all other loads are totaled,

$$26A + 38A + 19A + 19A + 5.7A + 8.3A + 11A = 127A$$

Applying the demand factor per Table 610.14(E), based on 6 cranes/hoists (motors) combinations* = 81 percent (.81).

$$127A \times .81 = 102.87A$$

— 141 —

This value is now added to the largest branch-circuit overcurrent device,

$$90A + 102.87A = 192.87A$$

A 175A overcurrent device must be used. Although, the calculated value (192.87A) exceeds the derived rating of the overcurrent device, the next standard size overcurrent device is not allowed per NEC 610.41(A).

*Combination motor loads per **Crane 2** and **Hoist 1** treated as single motors.

610.42(A) - Fuse or Circuit Breaker Rating (Branch-Circuit Short-Circuit and Ground-Fault Protection)

9. If dual element fuses were used to provide overcurrent protection for the motor in question No. 4. what size fuses would be required?

NEC 610.42(A) reference the use of Table 430.52 for selecting overcurrent protection for crane, hoist and monorail hoist motor branch circuits. Again, the motor's nameplate current rating is used for sizing the required overcurrent protection device.

For an *ac* 3ϕ squirrel-cage motor the motor's nameplate current is required to be increased by 175 percent (1.75). Therefore,

$$95A \times 1.75 = 166.25A$$

Exception No. 1 to NEC 430.52(C)(1) permits the use of the next higher standard overcurrent device, which in this case would be a set of 175A dual element fuses. See NEC 240.6(A).

610.53 - Overcurrent Protection

10. If 16 AWG conductors with a rated ampacity of 9 amps were used as the control circuits for a hoist motor. What size overcurrent device would be required to protect the conductors?

NEC 610.53 states that control circuits shall be considered as protected by overcurrent devices that are rated or set at not more than 300 percent (3) of the ampacity of the control conductors unless otherwise permitted.

Therefore, based on the ampacity of the 16 AWG conductors, where

$$9A \times 3 = 27A$$

The conductors are required to be protected by an overcurrent device that's rated for 25A. Using a 30A device would cause the conductors to be protected by an overcurrent device that exceeds 300 percent of the conductor's ampacity.

610.53(A) - Taps to Control Transformer (Overcurrent Protection)

11. What size overcurrent protection is required to protect the secondary circuit conductors that supply the control mechanism for a crane when tapped from a 480/120V single-phase 1kVA control transformer, where the ampacity of the secondary circuit conductors is 12A?

When the secondary circuit is protected by a device that's rated or set at not more than **(1)** 200 percent (2) of the rated secondary current of a control transformer *and* not more than **(2)** 200 percent (2) of the ampacity of the control circuit conductors, taps to control transformers are considered protected according to NEC 610.53(A).

Secondary current (I_S) of Control Transformer

$$\frac{1000VA}{120V} = 8.33A$$

Requirement (1) - At 200 percent of the rated secondary current

$$8.33A \times 2 = 16.66A$$

Overcurrent device must be rated for 15 amps. Using a 20 amp overcurrent device would exceed 200 percent.

Requirement (2) - 200 percent of the control circuit's conductor ampacity

$$12A \times 2 = 24A$$

Although a 20 amps overcurrent device would satisfy **Requirement (2)** that is, by not exceeding 200 percent of the control circuit's ampacity, it does on the other hand exceed **Requirement (1)**. Therefore, a 15 amps overcurrent device must be used in this situation because both requirements had to be met in order to use a 20 amps overcurrent device.

ARTICLE 620 - Elevators, Dumbwaiters, Escalators, Moving Walks, Platform Lifts, and Stairway Chair Lifts

> Refer to Worksheet G (Items 6C and 6D – MOTOR LOADS) (Volume 4) for applicable formulas pertaining to **Article 620**.

620.13(A) - Conductors Supplying Single Motor (Feeder and Branch-Circuit Conductors)

See questions relating to NEC 430.22 and 430.22(E)

620.13(B) - Conductors Supplying Single Motor Controller (Feeder and Branch-Circuit Conductors)

1. The nameplate current rating of a motor controller is listed as 23A. What size conductors are required to supply the controller and a 35A load?

Conductors supplying a single motor controller and other connected loads must have an ampacity not less than the combined loads. As a result,

$$23A + 35A = 58A$$

The ampacity of the supply conductors cannot be less than 58A.

620.13(C) - Conductors Supplying a Single Power Transformer (Feeder and Branch-Circuit Conductors)

2. The nameplate primary current rating of a transformer is 45A. Determine the ampacity of the conductors needed to supply the transformer and the following connected loads?

47A - continuous load   84A - non-continuous load

Along with the primary current and the connected loads, the supplying conductors must have an ampacity not less the combined loads. As a result,

$$45A + (47A \times 1.25) + 84A = 187.75A$$

620.13(D) - Conductors Supplying More than One Motor, Motor Controllers, or Power Transformer (Feeder and Branch-Circuit Conductors) (3. - 4.) [2]

3. The west wing of a hospital is equipped with eight elevators that are supplied by a 480V, three-phase feeder. Each elevator motor is driven by a motor-generator set that's rated for continuous duty. The horsepower rating and nameplate current (NPC) for each 460V, three-phase drive motor for each motor-generator set is as follows:

FREIGHT ELEVATOR 1
  75HP, NPC = 87A

FREIGHT ELEVATOR 2
  50HP, NPC = 58A

PATIENT ELEVATOR 1
30HP, NPC = 35A

PATIENT ELEVATOR 2
25HP, NPC = 28A

PASSENGER ELEVATORS 1 and 2
60HP, NPC = 70A

PASSENGER ELEVATORS 3 and 4
40HP, NPC = 47A

To operate doors, control mechanisms and other operating components, a 14A continuous rated controller is provided for each elevator. Based on the following information, determine the ampacity of the feeder conductors required to supply these loads.

NEC 620.13(D) reference the use of Table 430.22(E), NEC 430.24 and NEC 430.24, *Exception No. 1* for determining the ampere rating of those motors which will contribute to determining the ampacity of the conductors supplying the motors.

Motors used for elevators are classified for intermittent duty according to Table 430.22(E). NEC 430.24, *Exception No. 1* reference the use of NEC 430.22(E) for such motors when determining the ampacity of conductors supplying intermittent duty motors.

NEC 620.13 states that a conductor's ampacity must be based on the nameplate current rating of the driving motor of the motor-generator set where in this case all motor-generator sets are rated for continuous duty.

NEC 430.24 states that the conductors supplying several motors, or a motor(s) and other load(s), shall have an ampacity not less than **(1)** 125 percent(1.25) of the full-load current (FLC) rating of the highest rated motor *plus* **(2)** the sum of the full-load current ratings of other motors in the group (per motor's FLC tables) *plus* **(3)** 100 percent of the noncontinuous non-motor load(s) *plus* **(4)** 125 percent of the continuous non-motor load(s). All motor applications and current ratings must be in compliance with NEC 430.6(A).

NEC 430.24, *Exception No. 1* give reference for determining the highest rated motor. Referring to Table 430.22(E) per NEC 430.22(E), when a continuous rated motor is used for intermittent duty, 140 percent (1.40) of the motor's nameplate current (ampere rating) must be used. Observe,

75HP motor - (Largest FL and NP current ratings of all motors)

Per (Table 430.250) FLC = 96A   Per (NEC 430.24, *Exception No. 1*) NPC = 87A
96A x 1.25 = 120A   87A x 1.40 = 121.8A

As a result, the greater of either ampere rating is 121.8A which in this case identifies the highest rated motor of all motors.

AMPERE RATING OF HIGHEST RATED MOTOR -121.8A

AMPERE RATING OF OTHER MOTORS (Intermittent Duty - Continuous Rated)

$$([70A \times 2] + 58A + [47A \times 2] + 35A + 28A) \times 1.40 + 121.8A = 618.8A$$

Applying the Feeder Demand Factor per Table 620.14 (assumed - not under constant load)
8 elevators = .75 (75 percent)
618.8A x .75 = 464.1A

AMPACITY OF OTHER LOADS

14A [continuous rating] controller (8). Per NEC 215.2(A)(1) - (125 percent of continuous loads)

$$14A \times 8 \times 1.25 = 140A$$

TOTAL

$$464.1A + 140A = 604.1A$$

The ampacity of the feeder conductors required to supply the given loads must be no less than 604.1A.

4. Five 480V, three-phase transformers are used to supply an equal number of elevators and controllers. The transformers are rated for 45kVA and each controller has a nameplate current rating of 9A continuous. In addition, each elevator requires a non-continuous load of 3A to supply other components. Determine the ampacity of the feeder conductors required to supply these loads.

According to NEC 620.13(D) and Table 620.14

Determine Nameplate current rating of Transformers

$$I = \frac{45,000VA \ (45kVA)}{480V \times 1.732} = 54.13A$$

Total current based on 5 elevators

$$54.13A \times 5 = 270.65A$$

Applying the Feeder Demand Factor per Table 620.14 (assumed - not under constant load)

5 elevators = .82 (82 percent)
270.65A x .82 = 221.93A

Ampacity of Other Loads (5)

9A (continuous rating) controllers

3A (non-continuous rating) other components. Per NEC 215.2(A)(1) - (125 percent of continuous loads + 100 percent non-continuous loads)

$$(9A \times 5 \times 1.25) + (3A \times 5) = 71.25A$$

Total - $221.93A + 71.25A = 293.18A$

The ampacity of the feeder conductors required to supply the given loads must be no less than 293.18A.

620.61(B)(1) - Duty Rating on Elevator, Dumbwaiter, and Motor Generator Sets Driving Motors (Overload Protection for Motors)

5. A 460V, 3ϕ motor-generator (MG) set is used to drive an elevator motor. The MG set has a continuous rating and a nameplate current of 62.5A. What size overload protection is required for the motors?

According to NEC 620.61(B)(1) such motors must be rated as intermittent and protected against overload in accordance with NEC 430.33. NEC 430.6(A)(2) requires a motor's nameplate current to be used for sizing overload protection. NEC 430.33 states that motors used for intermittent duty per Table 430.22(E) are permitted to be protected against overload by its branch-circuit overcurrent device (fuse or circuit breaker) providing the rating or setting of the overcurrent device does not exceed that specified in Table 430.52.

Per Table 430.22(E), the nameplate current is increase by 140 percent (1.40) - (Intermittent duty/continuous rated).

$$62.5A \times 1.40 = 87.5A$$

Per Table 430.52,

if time-delay fuses are used, 87.5A is increased by 175 percent (1.75)

$$87.5A \times 1.75 = 153.13A$$

if an inverse-time circuit breaker is used, 87.5A is increased by 250 percent (2.50)

$$87.5A \times 2.50 = 218.75A$$

Because the rating or setting of either protective device cannot exceed that specified in Table 430.52 per NEC 430.33, 150A time-delay fuses or a 200A inverse-time circuit breaker can be used for overload protection.

620.61(B)(2) - Duty Rating on Escalator Motors (Overload Protection for Motors)

See related question Nos. 31. - 35. of Article 430.

620.61(B)(4) - Duty Rating and Overload Protection on Platform and Stairway Chair Lift Motors (Overload Protection for Motors)

See question No. 5.

ARTICLE 625 - Electric Vehicle Charging System

625.21 - Overcurrent Protection (Control and Protection)

Same as procedures given in NEC 210.20(A) and 215.3 for branch and feeder overcurrent protection (Volume 2). See question No. 8. of Article 210 and question Nos. 1. - 4. of Article 215.

625.29(D)(2) - Other Values (Ventilation Required) (1. - 2.) [2]

1. Determine the minimum ventilation requirements in cubic meters per minute (m^3/min) when a 25A branch circuit is used at 125V.

Using the formula given in NEC 625.29(D)(2)(1) for single phase circuits,

$$\text{Ventilation } (1\phi)(m^3/min) = \frac{125V \times 25A}{1718} = 1.82$$

The minimum ventilation requirement is 1.82 m^3/min.

2. Determine the minimum ventilation requirements in cubic feet per minute (cfm) when a three phase, 45A branch circuit is used at 465V.

Using the formula given in NEC 625.29(D)(2)(2) for three phase circuits,

$$\text{Ventilation } (3\phi)(cfm) = \frac{1.732 \times 465V \times 45A}{48.7} = 744.19$$

The minimum ventilation requirement is 744.19 cfm.

ARTICLE 626 - Electrified Truck Parking Spaces

626.11(A) - Parking Space Load (Feeder and Service Load Calculations)

1. At 240V, it was determined that the average single-phase load per parking space for an electrified truck would require 26A. If needed to size a feeder, calculate the kVA demand per parking space.

Based on the given information, the kVA demand per parking space results to 6.24kVA (240V x 26A/1000). However, the provisions of NEC 626.11(A) states that an electrical feeder shall be calculated on the basis of not less than 11kVA per electrified truck parking space. As a result, the 6.24kVA demand much be increased to 11kVA for sizing a feeder.

626.11(B) - Demand Factors (Feeder and Service Load Calculations)

2. Determine the minimum kVA feeder demand per Table 626.11(B) if the parking space per question No. 1. was in a "2b" climatic temperature zone.

Referring to the Table, a "2b" climatic temperature zone reference a 62 percent (.62) demand factor. Based on the demand factor, the minimum kVA feeder demand would result to 6.82kVA (11kVA x .62).

626.11(C) - Two or More Electrified Truck Parking Spaces (Feeder and Service Load Calculations)

3. Determine the minimum size electrical service needed at 208/120V-3ϕ to supply a 40 space parking lot for electrified trucks if located in a "5a" climatic temperature zone.

Where two or more electrified parking spaces are involved, NEC 626.11(C) states that the 11kVA minimum per NEC 626.11(A) must be applied per parking space. Applying the 47 percent (.47) demand factor per Table 626.11(B) based on the location of a "5a" climatic temperature zone and the 11kVA (11,000VA) minimum per parking space the electrical service is sized.

$$\frac{11,000VA \times 40 \times .47}{208V \times 1.732} = 574A$$

Based on the calculated results, as a minimum a 600A service is needed.

ARTICLE 630 - Electric Welders

> Refer to Worksheet G (Item 8A - INDUSTRIAL EQUIPMENT) (Volume 4) for applicable formulas pertaining to **Article 630**.

630.11(A) - Individual Welders (Arc Welders) (1. - 2.) [2]

1. A transformer arc welder operates at a 60% (.60) *duty cycle. Determine the minimum ampacity of the conductors required to supply the welder if the rated primary current of the welder is 55 amps.

*The operating capacity of a welder is basically determined by how much current it will generate at a given time period or duty cycle. Duty cycle is the number of minutes based on a 10 minute time period that a welder can operate. This means if a welder has a primary current rating of 400 amps and a 30% (.30) duty cycle it is capable of supplying 400 amps of current for 3 minutes (10 minutes x .30 = 3 minutes) and rest (cool down) for the remaining 7 minutes of the 10 minute time period to prevent overheating.

Because this is a transformer arc welder, the nonmotor generator column of Table 630.11(A) must be referenced to determine the multiplier for the welder at a duty cycle of 60% (.60). At 60%, the multiplier is .78 (78 percent) which is multiplied by the welder's primary current rating to determine the ampacity of the required conductors.

$$55A \times .78 = 42.9A$$

The minimum ampacity of the conductors must be 42.9A, the calculated supply current.

2. The nameplate of a motor-generator arc welder list the welder's primary current at 43 amps and a duty cycle of 80% (.80). If THWN copper conductors are used to supply the welder, what size conductors are required?

The motor-generator column of Table 630.11(A) lists a .91 (91 percent) multiplier for such welder when its duty cycle is 80 percent. Therefore,

$$43A \times .91 = 39.13A$$

Referring to Table 310.15(B)(16), an 8 AWG THWN copper conductor which has a temperature rating of 75°C and an ampacity of 50 amps is the minimum size conductor that can be used.

630.11(B) - Group of Welders (Arc Welders) (3. - 4.)[2]

3. Determine the minimum ampacity required for a single-feeder supplying six 240V motor-generator arc welders with a primary current rating of 27 amps and a 10% (.10) duty cycle.

NEC 630.11(B) references the provisions for determining the ampacity of a conductor supplying a group of arc welders. For a duty cycle of 20% or less, Table 630.11(A) lists a .55 (55 percent) multiplier for motor-generator arc welders. As a result, the arc welders supply current is 14.85A (27A x .55 = 14.85A).

After the multiplier is applied, NEC 630.11(B) states that the ampacity of a conductor supplying a group of welders is determined by summing the two largest welders at 100 percent (1), *plus* 85 percent (.85) of the third largest welder, *plus* 70 percent (.70) of the fourth largest welder, *plus* 60 percent (.60) of all remaining welders. Therefore,

$$
\begin{array}{ll}
(\mathbf{1} \text{ and } \mathbf{2}) \text{ - Two largest welder (@100 percent)} & - 14.85A \times 2 = 29.70A \\
(\mathbf{3}) \text{ - Third largest welder (@85 percent)} & - 14.85A \times .85 = 12.62A \\
(\mathbf{4}) \text{ - Fourth largest welder (@70 percent)} & - 14.85A \times .70 = 10.40A \\
(\mathbf{5} \text{ and } \mathbf{6}) \text{ - Remaining welders (@60 percent)} & - 14.85 \times .60 \times 2 = \underline{17.82A} \\
& 70.54A
\end{array}
$$

The minimum ampacity of the required feeder conductors is based on 70.54A, the total supply current of the motor-generator arc welders.

4. Determine the minimum ampacity required for a 240V, single-phase feeder supplying the following transformer arc welders:

	Primary Current Rating	Duty Cycle
Welder 1	36A	.80 (80 percent)
Welder 2	24A	.20 (20 percent)
Welder 3	60A	.50 (50 percent)
Welder 4	49A	.90 (90 percent)
Welder 5	24A	.50 (50 percent)
Welder 6	53A	.70 (70 percent)
Welder 7	18A	.30 (30 percent)

Per Table 630.11(A), first apply the appropriate multiplier per Table 630.11(A) per individual welder.

	Primary Current Rating	Duty Cycle	Multiplier
Welder 1	36A	.80 (80%)	.89
Welder 2	24A	.20 (20%)	.45
Welder 3	60A	.50 (50%)	.71
Welder 4	49A	.90 (90%)	.95
Welder 5	24A	.50 (50%)	.71
Welder 6	53A	.70 (70%)	.84
Welder 7	18A	.30 (30%)	.55

Calculate the minimum conductor ampacity per welder.

$$
\begin{array}{l}
\text{Welder 1 - } 36A \times .89 = 32.04A \\
\text{Welder 2 - } 24A \times .45 = 10.80A \\
\text{Welder 3 - } 60A \times .71 = 42.60A \\
\text{Welder 4 - } 49A \times .95 = 46.55A
\end{array}
$$

Welder 5 - 24A x .71 = 17.04A
Welder 6 - 53A x .84 = 44.52A
Welder 7 - 18A x .55 = 9.90A

Although Welders 3 and 6 are the two largest welders based on their rated primary current after deriving each welder's individual supply current, the largest welders are now based on the welders that produces the largest supply current, which are now Welders 4 and 6. Therefore,

(**4** and **6**) - Two largest welders (@100 percent) - 46.55A + 44.52A   = 91.07A
  (**3**) - Third largest welder (@85 percent)   - 42.6A x .85   = 36.21A
  (**1**) - Fourth largest welder (@70 percent) - 32.04A x .70   = 22.43A
(**5**, **2** and **7**) - Remaining welders (@60 percent) - (17.04A + 10.80A + 9.9A) x .60 = 22.64A
                                                               172.35A

The minimum ampacity of the required feeder conductors is based on 172.35 amps, the total supply current of the transformer arc welders.

630.12(A) - For Welders (Overcurrent Protection)

5. What size overcurrent protection is required, if the intentions are to solely protect each individual welder in question No. 4.?

Per NEC 630.12(A) each (arc) welder shall have an overcurrent protection rated or set not more than 200 percent (2) of the welder's maximum current (I_{max}). If the maximum current is not provided, the overcurrent protection must be rated or set at not more than 200 percent of the welder's rated primary current. However, the rated primary current is practically the same as a welder's maximum current. NEC 630.12 permits the next standard size overcurrent device to be used if the calculated value does not correspond to a standard ampere rating per NEC 240.6 *or* where the rating or setting specified results in unnecessary opening (nuisance tripping) of the overcurrent device. Applying the given primary currents:

	Primary Current Rating					Required Overcurrent Protection
Welder 1	36A	x	2	=	72A	80A
Welder 2	24A	x	2	=	48A	50A
Welder 3	60A	x	2	=	120A	125A
Welder 4	49A	x	2	=	98A	100A
Welder 5	24A	x	2	=	48A	50A
Welder 6	53A	x	2	=	106A	110A
Welder 7	18A	x	2	=	36A	40A

The selected overcurrent devices reflects the individual overcurrent protection required per welder for the primary purpose of protecting the welder opposed to the welder being protected by an overcurrent device whose primary purpose is to protect the conductors supplying the welder. If the overcurrent device is rated or set at the limits described in NEC 630.12(A), separate overcurrent devices are not required to protect the supplying conductors and the welder.

630.12(B) - For Conductors (Overcurrent Protection) (6. - 8.) [3]

6. If individual branch-circuit conductors were used to supply the welder's in question No. 4, what size overcurrent protection would be required to protect the conductors? Assume the branch-circuit conductors will be THWN copper.

Unlike the overcurrent protection sized in question No. 5., the overcurrent protection desired in question No. 6. pertains to protecting the conductors opposed to protecting the welders. Notice when an overcurrent device is used to protect a welder no more than 200 percent of the welder's primary current is applied towards selecting the overcurrent device. However, when an overcurrent device is used to protect the conductors supplying the welders no more than 200 percent of the ampacity of the conductors supplying a welder or a group of welders is applied towards selecting the overcurrent device. Here, the ampacity of the conductors does not mean the minimum calculated ampacity but the ampacity of the selected conductors [Table 310.15(B)(16)] resulting from the minimum calculated ampacity. In question No. 4., for each welder the minimum ampacity is the product of the welder's rated primary current and duty cycle multiplication factor.

	Calculated Ampacity	Conductor* Amapcity					Required Overcurrent Protection
Welder 1	32.04A	(10 AWG) 35A	x	2	=	70A	70A
Welder 2	10.80A	(14 AWG) 20A	x	2	=	40A	40A
Welder 3	42.60A	(8 AWG) 50A	x	2	=	100A	100A
Welder 4	46.55A	(8 AWG) 50A	x	2	=	100A	100A
Welder 5	17.04A	(14 AWG) 20A	x	2	=	40A	40A
Welder 6	44.52A	(8 AWG) 50A	x	2	=	100A	100A
Welder 7	9.90A	(14 AWG) 20A	x	2	=	40A	40A

*[Per Table 310.15(B)(16) – THWN Copper]

With the exception of two overcurrent devices (underlined) notice the differences between the overcurrent protection based on the welders being protected opposed to the conductors.

7. What size overcurrent protection is required to protect the feeder conductors in question No. 3?

Based on the minimum ampacity of the required feeder conductors, 70.54A, NEC 630.12(B) states, "for conductors that supply one or more welders" where, in this case the minimum ampacity of the feeder conductors is based on six welders opposed to the individual overcurrent protection provided for each welder in question No. 6.

Being the case, per Table 310.15(B)(16) where copper conductors at a temperature rating of 75°C is applied, 4 AWG conductors are required which has an ampacity of 85A. Because the overcurrent device required to protect the feeder conductors must be rated or set at not more than 200 percent (2) of the conductor's ampacity such requirement results to,

$$85A \times 2 = 170A$$

where, a 150A overcurrent device is needed to serve as the protective device for the feeder conductors. The use of a 175A overcurrent protective device could not be used because this device would exceed the 200 percent limitation.

8. What size overcurrent protection is required to protect the feeder conductors in question No. 4?

Following the procedure used in question No. 7 and applying the minimum ampacity of the required feeder conductors based on 172.35 amps, per Table 310.15(B)(16) where copper conductors at a temperature rating of 75°C is again applied, 2/0 AWG copper conductors are required which have a rated ampacity of 175A. Applying the 200 percent (2) limitation, the results amounts to,

175A x 2 = 350A

where, a 350A overcurrent device is needed to serve as the protective device for the feeder conductors.

630.13 - Disconnecting Means (Overcurrent Protection)

When applied, the rating of a disconnecting means supporting an individual welder or a group of welder must accommodate the overcurrent protection of either the welder(s) or the conductors supplying the welder(s).

630.31(A)(1) - Individual Welders (Resistance Welders - Ampacity of Supply Conductors)

9. What size conductors are needed per NEC 630.31(A)(1) to supply the following resistance welders:

Automatically fed – Primary current - 58 amps
Nonautomatic – Primary current - 63 amps

According to NEC 630.31(A)(1) the ampacity of conductors supplying an automatically fed resistance welder must not be less than 70% (.70) of the welder's rated primary current and not less than 50% (.50) of a nonautomatic resistance welder's primary current. NEC 630.31(A)(1) makes reference to resistance welders *that may be operated at different times at different values* of its primary current or duty cycle when considering the required ampacity of the welder's supply conductors. To determine such ampacities observe the following calculations,

Automatically fed welder - 58A x .70 = 40.6A
Nonautomatic welder - 63A x .50 = 31.5A

Based on the calculations, the conductors supplying the automatically fed welder must have an ampacity no less than 40.6A and the conductors supplying the nonautomatic welder must have an ampacity no less than 31.5A.

630.31(A)(2) - Individual Welders (Resistance Welders - Ampacity of Supply Conductors) (10. - 11.) [2]

10. Determine the minimum ampacity required to supply each individual welder per NEC 630.31(A)(2) based on a 50% (.50) duty cycle and the given primary currents:

Welder (1) - 92.6 amps
Welder (2) - 77 amps
Welder (3) - 68.7 amps

The minimum ampacity required to supply each individual welder is determined per specified multiplier (.71 @ 50% duty cycle) listed in Table 630.31(A)(2).

Welder (**1**) - 92.6A x .71 = 65.75A
Welder (**2**) - 77A x .71 = 54.67A
Welder (**3**) - 68.7A x .71 = 48.78A

Opposed to NEC 630.31(A)(1), NEC 630.31(A)(2) makes reference to resistance welders *that are wired for a specific operation* based on the welder's primary current and duty cycle when considering the required ampacity of the welder's supply conductors.

11. What size THW copper conductors are needed to supply a 480V, three-phase resistance welder having a constant 40% (.40) duty cycle and a 72.3kVA operating load?

To begin, let's determine the constant operating load amps of the welder,

$$I = \frac{72.3\text{kVA}}{480\text{V} \times 1.732} = 86.97\text{A}$$

According to NEC 630.31(A)(2), the ampacity of conductors supplying a resistance welder when the welder's duty cycle and primary current are unchanged (constant) requires the actual (operating) primary current and the multiplier listed in Table 630.31(A)(2) to be applied to determine the ampacity of the supply conductors. For a 40% duty cycle, the multiplier .63 is used. With a constant operating load of 86.97A where,

86.97A x .63 = 54.79A

The results of the calculation require the ampacity of the supply conductors to be no less than 54.79A. Therefore as a minimum, per Table 310.15(B)(16), 6 AWG THW copper conductors are required.

630.31(B) - Group of Welders (Resistance Welders - Ampacity of Supply Conductors)

12. What size feeder conductors are required to supply the following resistance welders?

Welder (1) - Automatically fed – Primary current - 22 amps
Welder (2) - Seam – Primary Current - 31 amps
Welder (3) - Nonautomatic – Primary current - 43 amps
Welder (4) - 13.7kVA load with 30% duty cycle (constant)
Welder (5) - 11.4kVA load with 25% duty cycle (constant)
Welder (6) - 9.35kVA load with 7.5% duty cycle (constant)

The feeder conductors will be supplied by a 240V-1φ source. Use 75°C copper conductors.

Referring to NEC 630.31(A)(1) and (2)

Welders (**1**) and (**2**) (Automatically fed and seam welder) - 70 percent (.70) of primary current

$$22A \times .70 = 15.4A$$
$$31A \times .70 = 21.7A$$

Welder (**3**) (Nonautomatic welder) - 50 percent (.50) of primary current

$$43A \times .50 = 21.5A$$

Welders (**4**) - (**6**) (Constant Operation)

$$I_4 = \frac{13.7kVA}{240V} = 57.08A \qquad I_5 = \frac{11.4kVA}{240V} = 47.5A \qquad I_6 = \frac{9.35kVA}{240V} = 38.96A$$

Welder (**4**) 50% duty cycle = .71 multiplier
Welder (**5**) 25% duty cycle = .50 multiplier
Welder (**6**) 7.5% duty cycle = .27 multiplier

(**4**) 57.08A x .71 = 40.53A
(**5**) 47.5A x .50 = 23.75A
(**6**) 38.96A x .27 = 10.52A

Based on the given primary currents of Welders (**1**) - (**3**) and the calculated operational currents for Welders (**4**) - (**6**), Welder (**4**) is identified as the largest welder. According to NEC 630.31(B), the ampacity of conductors that supply a group of welders is based on the largest welder per calculated ampacity at 100 percent and at 60 percent (.60) of the calculated ampacities for all other welders. As a result, the ampacity of the feeder conductors are determined as such,

$$40.53A + [(15.4A + 21.7A + 21.5A + 23.75A + 10.52A) \times .60] = 96.25A$$
Welders   (**4**)   (**1**)   (**2**)   (**3**)   (**5**)   (**6**)

The welder's feeder conductors must have an ampacity no less than 96.25A. Referring to Table 310.15(B)(16) as a minimum, 3 AWG copper conductors rated for 75°C are required.

630.32(A) - For Welders (Overcurrent Protection for Resistance Welders) (13. - 14.) [2]

13. What size overcurrent devices are required to protect the welders in question Nos. 10. and 12.?

NEC 630.32(A) states that the overcurrent device for a resistance welder must be rated or set at not more than 300 percent (3) of the rated primary current. As a reminder, rated or set at not more than 300 percent of the rated primary current means the rating of the overcurrent device cannot be increased to the next higher standard size overcurrent device. However, NEC 630.32 permits the next standard size overcurrent device to be used if the calculated value does not correspond to a standard ampere rating per NEC 240.6(A) *or* where the rating or setting specified results in unnecessary opening (nuisance tripping) of the overcurrent device.

Question No. 10.

	Primary Current Rating					Required Overcurrent Protection
Welder 1	92.6A	x	3	=	277.8A	300A
Welder 2	77A	x	3	=	231A	250A
Welder 3	68.7A	x	3	=	206.1A	225A

Question No. 12.

	Primary Current Rating					Required Overcurrent Protection
Welder 1	22A	x	3	=	66A	70A
Welder 2	31A	x	3	=	93A	100A
Welder 3	43A	x	3	=	129A	150A
Welder 4	57.08A	x	3	=	171.24A	175A
Welder 5	47.5A	x	3	=	142.5A	150A
Welder 6	38.96A	x	3	=	116.88A	125A

14. What size overcurrent devices are required to protect the supply conductors in question Nos. 10. and 12. per NEC 630.32(A).?

NEC 630.32(A) states if the supply conductors for a welder are protected by an overcurrent device rated or set at not more than 200 percent (2) of the rated primary current of the welder, a separate overcurrent device shall not be required to protect the welder. Again, NEC 630.32 permits the next standard size overcurrent device to be used if the calculated value does not correspond to a standard ampere rating per NEC 240.6(A) *or* where the rating or setting specified results in unnecessary opening (nuisance tripping) of the overcurrent device.

Question No. 10.

	Primary Current Rating					Required Overcurrent Protection
Welder 1	92.6A	x	2	=	185.2A	200A
Welder 2	77A	x	2	=	154A	200A
Welder 3	68.7A	x	2	=	137.4A	150A

Question No. 12.

	Primary Current Rating					**Required Overcurrent Protection**
Welder 1	22A	x	2	=	44A	50A
Welder 2	31A	x	2	=	62A	70A
Welder 3	43A	x	2	=	86A	90A
Welder 4	57.08A	x	2	=	114.16A	125A
Welder 5	47.5A	x	2	=	95A	100A
Welder 6	38.96A	x	2	=	77.92A	80A

According to NEC 630.32(A), a separate overcurrent device is not required for corresponding welders and conductors, if the overcurrent device is rated or set at no more than 200 percent of the rated primary current of the welder.

630.32(B) - For Conductors (Overcurrent Protection for Resistance Welders)

15. What size overcurrent devices are required to protect the conductors in question Nos. 11. and 12.?

The 6 AWG THW copper conductors required in question No. 11. have a 65A conductor rating. NEC 630.32(B) requires the conductors to be protected by an overcurrent device rated for or set no more than 300 percent (3) of the conductor's rating. However, NEC 630.32 allows otherwise. As a result,

$$65A \times 3 = 195A$$

A 200A overcurrent device can be used.

The 3 AWG copper conductors required in question No. 12. have a 100A conductor ampacity. Again, these conductors are required to be protected by an overcurrent device rated for or set no more than 300 percent (3) of the conductor's ampacity. Again, NEC 630.32 allows otherwise. As a result,

$$100A \times 3 = 300A$$

A 300A overcurrent device can be used.

ARTICLE 645 - Information Technology Equipment

645.5(A) - Branch-Circuit Conductors (Supply Circuits and Interconnecting Cables)

1. Refer to question No. 91. in Article 220 (Volume 1). What size branch-circuit conductors and overcurrent protection are required to supply each main frame computer? What size branch-circuit conductors and overcurrent protection are required to supply both main frame computers? Assume the use of copper conductors at 75°C.

The two main frame computers are each rated for 17,500VA at 208V-3ϕ. With this information the computer's operating currents can be determined where,

$$\frac{17,500VA}{208V \times 1.732} = 48.58A$$

Now that the main frame computer's operating currents have been determined, NEC 645.5(A) requires the branch-circuit conductors supplying one or more units of a data processing system to have an ampacity not less than 125 percent (1.25) of the total connected load. This 125 percent requirement will satisfy the continuous operating conditions of the computers.

Individual Branch-circuit Conductors and Overcurrent Protection

The ampacity of the conductors needed to supply each main frame computer amounts to:

$$48.58A \times 1.25 = 60.73A$$

At 60.73A, per Table 310.15(B)(16), 6 AWG copper conductors are required to supply each main frame computer. At 75°C, 6 AWG copper conductors have a rated ampacity of 65A. Per NEC 240.4(B), a 70A overcurrent device can be used to protect each computer.

Total Load Branch-Circuit Conductors and Overcurrent Protection

The ampacity of the conductors needed to supply both main frame computers (the total connected load) amounts to:

$$48.58A \times 2 \times 1.25 = 121.45A$$

At 121.45A, per Table 310.15(B)(16), 1 AWG copper conductors are required to supply both main frame computers. At 75°C, 1 AWG copper conductors have a rated ampacity of 130A. A 125A overcurrent device can be used to protect both computers. However, per NEC 240.4(B), a 150A overcurrent device can be used as well.

ARTICLE 647 - Sensitive Electronic Equipment

647.4(D)(2) - Cord-Connected Equipment (Voltage Drop)

1. A 120V, 15A branch circuit originating in a dedicated panelboard supplies a group of receptacles that serves sensitive electronic equipment. The distance from the panelboard to the receptacles is 53'. What size copper conductors are required to supply the receptacles to maintain a 1 percent voltage drop?

The voltage drop on branch circuits supplying receptacles shall not exceed 1 percent (.01) according to NEC 647.4(D)(2). NEC 647.4(D)(2) also states that the load connected to the receptacle outlet when making such (voltage drop) calculations shall be 50 percent of the branch-circuit rating, which in this problem would be 7.5A. Using the converted voltage drop formula,

$$\mathbf{R} = \frac{V_D \text{ x } 1000'}{2 \text{ x L x I}}$$

where **R** is equal to the resistance of the conductor needed to maintain a 1 percent voltage drop and V_D = 1.2V (120V x .01) results to,

$$\mathbf{R} = \frac{1.2 \text{ x } 1000'}{2 \text{ x } 53' \text{ x } 7.5A} = 1.51\Omega$$

Referring to Table 8 of Chapter 9, the first uncoated copper conductor which has a resistance less than 1.51Ω is a 10 AWG conductor which has a resistance of 1.21 ohms/kFt for solid copper conductors and 1.24 ohms/kFt for stranded copper conductors. As a minimum, a 10 AWG copper conductor is required to maintain a 1 percent voltage drop for conductors supplying receptacles that serve cord-connected equipment.

ARTICLE 660 - X-Ray Equipment

660.5 - Disconnecting Means

1. Refer to question No. 1. of Article 517 [NEC 517.72(A)]. The only difference between the requirements for the electrical installation of X-ray equipment referenced in Article 660 compared to that of Article 517, is Article 660 reference the electrical installation of X-ray equipment for nonmedical and nondental use and Article 517 reference the electrical installation of X-ray equipment for medical and dental use. Based on the calculated results of the X-ray unit in question No. 1. of Article 517, what size disconnecting means would be required based on the provisions of NEC 660.5?

Again, based on the calculated determinant per question No. 1. of Article 517, a disconnecting means with a standard rating exceeding 31.3A would be required.

660.6(A) - Branch-Circuit Conductors (Rating of Supply Conductors and Overcurrent Protection)

Same application as 517.73(A)(1) of Article 517, see question No. 2.

660.6(B) - Feeder Conductors (Rating of Supply Conductors and Overcurrent Protection)

Similar application as 517.73(A)(2) of Article 517, see question No. 3.

ARTICLE 665 - Induction and Dielectric Heating Equipment

665.10(A) - Nameplate Rating (Ampacity of Supply Conductors)

1. A certain type three-phase industrial machine is rated for 460V. According to the machine's nameplate, two related pieces of equipment consist of synchronized loads that pull 34 amps and 22 amps each. The machine's nameplate also consists of two other non-synchronized heating and melting loads that pulls a total of 41 amps. Determine the minimum ampacity of the conductors needed to supply the machine. Based on the ampacity of the conductors, what size 90°C RHW-2 copper or aluminum conductors are needed?

Assuming all requirements of NEC 665.10(A) are met, the minimum ampacity of the conductors needed to supply the machine totals 97A (34A + 22A [simultaneous operation] + 41A [remaining machines nameplate ratings]). Based on the derived ampacity, per Table 310.15(B)(16), 3 AWG copper (115A) or 2 AWG aluminum (100) conductors are needed.

ARTICLE 669 - Electroplating

669.5 - Branch-Circuit Conductors

1. A three-phase, 208V branch-circuit supplies two pieces of equipment that's used for electroplating and electropolishing. The electroplating equipment has a rated load of 21kVA while the electropolishing equipment is rated for 17kVA. Determine the ampacity of the branch-circuit conductors needed to supply both pieces of equipment.

Applying the rated loads of the equipment, the equipment's operating currents are determined where,

$$\frac{21,000VA}{208V \times 1.732} = 58.29A \qquad \frac{17,000VA}{208V \times 1.732} = 47.19A$$

NEC 669.5 requires the branch-circuit conductors supplying the equipment to have an ampacity not less than 125 percent (1.25) of the total connected load. Based on the total connected load of the equipment, the ampacity of the branch-circuit conductors needed to supply the total load amounts to 131.85A ([58.29A + 47.19A] x 1.25).

For questions pertaining to the ampacities of busbars, refer to NEC 366.23(A) questions 6. and 7. of Article 366 (Volume 2).

ARTICLE 670 - Industrial Machinery

670.4(A) - Size (Supply Conductors and Overcurrent Protection)

1. An industrial machine will be supplied by a three-phase 480V feeder. The machine consists of the following loads:

Lamination Heating - 59A   Electrical Press - 38A
40HP motor (continuous duty)   25HP motor (2) (continuous duty)
10HP motor (short-time duty-5 minute rated)   7.5HP motor (2) (varying duty-15 minute rated)

What size conductors are required to supply the machine? Assume the use of copper conductors at 75°C.

In sizing the supply conductors for the industrial machine, NEC 670.4(A) requires the ampacity of such conductors to be no less 125 percent (1.25) of the full-load current rating of **(1)** all resistance heating loads *plus* 125 percent (1.25) of the full-load current rating of **(2)** the highest rated motor *plus* **(3)** the sum of the full-load current ratings of all other connected motors and **(4)** apparatus, based on their duty cycle, that may be in operation at the same time.

To determine the ampacity of the supply conductors for this application the given loads are categorized per instructions. All loads are expected to operate at some given time simultaneous.

(1) Resistance Heating Loads = 59A
(2) Highest Rated Motor [per NEC 430.17 and Table 430.250] = 52A (40HP motor)
(3) Full-load current ratings of other motors [per Tables 430.22(E) and 430.250]
      25HP motor (2) (continuous duty) - 34A x 2 = 68A
      10HP motor (short-time duty - 5 minute rated)* - 14A x 1.10 = 15.4A
      7.5HP motor (2) (varying duty -15 minute rated)* - 11A x 1.20 x 2 = 26.4A
(4) All other apparatus
      Electrical Press = 38A

*Per NEC 430.22(E) such motors per duty applications reference the use of the motor's nameplate current. However, since the nameplate currents are not given, the applicable full-load currents of Table 430.250 are acceptable.

Applying the current ratings of the given loads per NEC 670.4(A) the required ampacity of the supply conductors is derived in order to determine the size of the supply conductors.

    (1) 59A x 1.25 + **(2)** 52A x 1.25 + **(3)** 68A + 15.4A + 26.4A + **(4)** 38A = 286.55A
      [73.75A]   **[65A]**   **[109.8A]**   **[38A]**

At 286.55A, per Table 310.15(B)(16) as a minimum, 350 kcmil copper conductors rated for 310A at 75°C are required to supply the industrial machine.

670.4(C) - Overcurrent Protection (Supply Conductors and Overcurrent Protection)

2. What size inverse-time circuit breaker or time-delay fuses are required to protect the supply conductors in question No. 1?

In sizing the overcurrent protection for the conductors supplying the industrial machine, NEC 670.4(C) requires **(1)** the overcurrent device to not be greater than the sum of the largest rating or setting of the protective device provided for the machine *plus* **(2)** 125 percent (1.25) of the full-load current rating of all resistance heating loads *plus* **(3)** the sum of the full-load current ratings of all other connected motors and **(4)** apparatus that could be in operation at the same time.

In compliance with the requirements of NEC 670.4(C), because the question does not state that a protective device is provided with the machine, the rating or setting of the overcurrent protective device must be determined based on 430.52 and 450.53, as applicable.

Where the use of either an inverse-time circuit breaker or time-delay fuses are used as the overcurrent protection, the provisions of NEC 430.53(C) are applied to determine the size of the overcurrent device that was not provided with the machine. It is assumed that all conditions of NEC 430.53(C) are met.

If an inverse-time circuit breaker or time-delay fuses are used in accordance with NEC 430.52 (based on the highest rated motor connected – 40HP motor at 52A), the inverse-time circuit breaker would be rated for 125A (52A x 2.50 = 130A) and the time-delay fuses (52A x 1.75 = 91) would be rated for 90A. The inverse time circuit breaker or time-delay fuses must have a rating not exceeding that specified in NEC 430.52 according to NEC 430.53(C)(4).

Per NEC 430.53(C)(4), the rating or setting of the overcurrent device for the circuit supplying the machine is determined by either of the following calculations:

$$\boxed{125A} + 59A + 68A + 15.4A + 26.4A + 38A = 331.8A \text{ (inverse-time circuit breaker)}$$
$$or$$
$$\boxed{90A} + 59A + 68A + 15.4A + 26.4A + 38A = 296.8A \text{ (time-delay fuses)}$$

In compliance with the last sentence of NEC 430.53(C)(4), if either device was used they could be rated no higher than 350A per NEC 240.4(B) based on the ampacity rating (310A) of the 350 kcmil copper supply conductors.

Now to apply the provisions of 670.4(C), let's assume the branch-circuit short-circuit and ground–fault protective device provided with the machine was either a 125A inverse-time circuit breaker or 90A time-delay fuses.

Per NEC 670.4(C), the overcurrent protective device for the circuit supplying the machine shall not be greater than the sum of the **(1)** largest rating or setting of the branch-circuit short-circuit and ground-fault protective device provided with the machine, *plus* **(2)** 125 percent (1.25) of the full-load current rating of all resistance heating loads, *plus* **(3)** the sum of the full-load currents of all other motors and **(4)** apparatus that could be in operation at the same time.

Applying both overcurrent device and the current ratings of the given loads per NEC 670.4(C), the required overcurrent protective device supplying the machine is derived.

Using a circuit breaker
    (1) 125A + **(2)** 59A x 1.25 + **(3)** 52A + 34 x 2 + 14A + 11A x 2 + **(4)** 38A = 392.75A
                 [73.75A]   **[68A]**   **[22A]**

Using fuses
    (1) 90A + **(2)** 59A x 1.25 + **(3)** 52A + 34 x 2 + 14A + 11A x 2 + **(4)** 38A = 357.75A
                 [73.75A]   **[68A]**   **[22A]**

Based on the derived results, whether an inverse-time circuit breaker or fuses are used the selected device cannot be sized larger than 350A also. Per NEC 670.4(C), the selected overcurrent device cannot be greater than the summation (392.75A or 357.75A) of the machine's protective device and all loads considered.

ARTICLE 675 - Electrically Driven or Controlled Irrigation Machines

675.7(A) - Continuous-Current Rating (Equivalent Current Ratings)

1. An irrigation machine is maneuvered by seven 1 HP motors and seven 1½ HP motors. The motors are three phase-240 volts and listed for intermittent duty with a 30-minute rating. The 1 HP motors have a nameplate full-load current rating of 3.4A and the 1½ HP motors have a nameplate full-load current rating of 5.3A. What size copper THW branch-circuit conductors are required to supply the irrigation machine? What size overcurrent protection is required?

According to NEC 675.7(A) the branch-circuit conductors and overcurrent protection shall be equal to **(1)** 125 percent (1.25) of the largest motor nameplate full-load current rating *plus* **(2)** the sum of the nameplate full-load current ratings of all remaining motors on the circuit *multiplied by* **(3)** the maximum percent duty cycle at which the motors can continuously operate.

Since the 1½ HP motors are the largest motors, one of the motors will be used as the largest motor. Along with the largest motor nameplate full-load current rating, the application of the remaining motor's nameplate full-load current ratings and the required duty cycle percentage at 90 percent (.90) per Table 430.22(E), the irrigation machine's branch-circuit conductors can be determined.

As it relates to NEC 675.7(A), the calculated ampacity of the branch-circuit is:

$$5.3A \times 1.25 \ [6.63A] + \{(5.3A \times 6 \ [31.8A] + 3.4A \times 7 \ [23.8A]) \times .90\} = 56.67A$$

Based on the calculated ampacity, 6 AWG THW copper conductors rated for 75°C per Table 310.15(B)(16) can be used as the branch-circuit conductors for the irrigation machine. A 60A overcurrent device is required to protect the irrigation machine.

675.7(B) - Locked-Rotor Current (Equivalent Current Ratings)

2. Calculate the motor's total locked-rotor current in question No. 1., if the motor's nameplate display the code letter E.

Applying the provisions of NEC 675.7(B), the equivalent locked-rotor current rating shall be equal to the numerical sum of the locked-rotor current of **(1)** the two largest motors *plus* **(2)** 100 percent of the sum of the motor nameplate full-load current ratings of all remaining motors on the circuit.

Applying the locked-rotor indicating code letter per Table 430.7(B) at worst case and the applicable formula, the locked-rotor current of the largest motor at 1½ HP is derived:

$$LRC_{(1½ HP)} = \frac{4.99kVA \times 1000 \times 1.5HP}{240V \times 1.732} = 18A$$

Per NEC 675.7(B), the equivalent locked-rotor current rating for the irrigation machine is so derived:

(1) 18A x 2 [36A] + **(2)** (5.3A x 5 [26.5A] + 3.4A x 7 [23.8A]) = 86.3A

The equivalent locked-rotor current rating for the irrigation machine is 86.3A.

675.11(A) - Transmitting Current for Power Purposes (Collector Rings)

Refer to NEC 675.7(A) for a similar application based upon the use of motors.

675.11(B) - Control and Signal Purposes (Collector Rings)

Application based on current ratings of devices. The 125 percent (1.25) increase to be applied in the same manner as an application involving a continuous load.

675.22(A) - Continuous-Current Rating (Equivalent Current Ratings)

3. If the irrigation machine in question No 1. was a center pivot irrigation machine, what size copper THW branch-circuit conductors and controller are required to serve the irrigation machine?

According to NEC 675.22(A), the branch-circuit conductors and overcurrent protection shall be equal to **(1)** 125 percent (1.25) of the largest motor nameplate full-load current rating *plus* **(2)** 60 percent (.60) of the sum of the nameplate full-load current ratings of all remaining motors on the circuit.

Again, since the 1½ HP motors are the largest motors, one of the motors will be used as the largest motor. Along with the largest motor nameplate full-load current rating and the application of the remaining motor's nameplate full-load current ratings calculated at 60 percent (.60), the center pivot irrigation machine's branch-circuit conductors can be determined.

As it relates to NEC 675.22(A), the calculated ampacity of the branch-circuit is:

$$5.3A \times 1.25\ [6.63A] + \{(5.3A \times 6\ [31.8A] + 3.4A \times 7\ [23.8A]) \times .60\} = 39.99A$$

Based on the calculated ampacity, 8 AWG THW copper conductors rated for 75°C per Table 310.15(B)(16) can be used as the branch-circuit conductors for the irrigation machine. As a minimum, a 40A overcurrent device is required to protect the center pivot irrigation machine.

675.22(B) - Locked-Rotor Current (Equivalent Current Ratings)

4.   Calculate the locked-rotor current in question No. 1. based on the same requirements as question No. 2. if a center pivot irrigation machine is used instead.

Applying the provisions of NEC 675.22(B), the equivalent locked-rotor current rating shall be equal to the numerical sum of the locked-rotor current of **(1)** the two largest motors *plus* **(2)** 80 percent (.80) of the sum of the motor nameplate full-load current ratings of all remaining motors on the circuit.

Applying the locked-rotor indicating code letter per Table 430.7(B) at worst case and the applicable formula, the locked-rotor current of the largest motor at 1½ HP is derived:

$$\text{LRC}_{(1\frac{1}{2}\text{ HP})} = \frac{4.99\text{kVA} \times 1000 \times 1.5\text{HP}}{240\text{V} \times 1.732} = 18A$$

Per NEC 675.22(B), the equivalent locked-rotor current rating for the irrigation machine is so derived:

$$\textbf{(1)}\ 18A \times 2\ [36A] + \textbf{(2)}\ \{(5.3A \times 5\ [26.5A] + 3.4A \times 7\ [23.8A]) \times .80\} = 76.24A$$

The equivalent locked-rotor current rating for the irrigation machine is 76.24A.

ARTICLE 680 - Swimming Pools, Fountains, and Similar Installations

680.9 - Electric Pool Water Heaters

Refer to Article 424. See question No. 3. for similar approach.

ARTICLE 690 - Solar Photovoltaic Systems

Table 690.7 -Voltage Correction Factors for Crystalline and Multicrystalline Silicon Modules

1.   Determine the correction factors for crystalline or multicrystalline silicon photovoltaic module where the ambient temperature is 56°F, 4°C, 13°F and -32°C.

The correction factors of Table 690.7 apply to ambient temperatures below 25°C or 77°F.

Ambient Temperature(s)	Correction Factor
56°F	1.06
4°C	1.10
13°F	1.16
-32°C	1.23

690.7(A) - Maximum Photovoltaic System Voltage (Maximum Voltage)

2.   The rated open circuit voltage for crystalline silicon modules is 35V. Determine the module's maximum voltage if placed in an environment where the ambient temperature is -8°C.

NEC 690.7(A) states for crystalline and multicrystalline silicon modules, the rated open-circuit voltage shall be multiplied by the correction factor provided in Table 690.7 to determine the module's maximum voltage. At -8°C, Table 690.7(A) reference a correction factor of 1.14. Therefore the modules maximum voltage is,

$$35V \times 1.14 = 39.9V$$

690.8(A)(1) - Photovoltaic Source Circuit Currents (Calculation of Maximum Circuit Current)

3. A photovoltaic panel consists of 72 modules. Supposing each module has a rated short-circuit current of 7.5A, determine the maximum photovoltaic source circuit current.

To provide a brief understanding of a photovoltaic panel, let's consider how the panel is constructed (refer to the definitions in NEC 690.2). A photovoltaic cell also referred to as a solar cell is a photovoltaic device that generates electricity when exposed to light. Just as in a series circuit where individual voltage drops are added together to produce a source voltage, individual photovoltaic cells are connected in series until a desired output voltage is obtained. These photovoltaic cells are then assembled together to form a module. A module or a combination of modules is called a photovoltaic source circuit which produces output current. Combinations of photovoltaic source circuits are joined together in parallel until a desired power rating has been reached. Once a series of modules are interconnected they form a panel.

NEC 690.8(A)(1) states that the maximum current shall be the sum of the parallel module's rated short-circuit currents multiplied by 125 percent. Based on this and the given information the sum of the 72 module's short-circuit current is,

$$72 \times 7.5A = 540A$$

The calculated product is then increased by 125 percent to derive the maximum photovoltaic source circuit current,

$$540A \times 1.25 = 675A$$

Although the panel's actual source circuit current will be far less than the maximum photovoltaic source circuit current, the maximum photovoltaic source circuit current is used to size circuit conductors to handle short-circuit current in the event of a short-circuit which could last for an extended period of time.

690.8(A)(2) - Photovoltaic Output Circuit Currents (Calculation of Maximum Circuit Current)

4. Determine the maximum output circuit current if an identical photovoltaic panel was added to the panel in question No. 3.

If an identical panel was connected to the existing panel they would be connected together in parallel. Just as in a parallel circuit, individual current values are added together to obtain a source current which is what would happen in this situation. In actuality this is what NEC 690.8(A)(2) is directing the user to do. Based on this, the maximum output circuit current would be twice the value calculated in question No. 3. which is,

$$675A \times 2 = 1350A$$

690.8(A)(3) - Inverter Output Circuit Current (Calculation of Maximum Circuit Current)

5. NEC 690.8(A)(3) states "the maximum current shall be the inverter continuous output current rating", explain this.

A *dc*-to-*ac* inverter is used to convert a direct current (*dc*) input into an alternating current (*ac*) output in a solar photovoltaic system. The output *ac* source of an inverter is connected to the load the solar photovoltaic system is supplying. This load just as others is probably based on a calculation. However, the maximum current supplying the load must not be based on load calculation's but instead on the continuous rated output current of the inverter.

690.8(A)(4) - Stand-Alone Inverter Input Circuit Current (Calculation of Maximum Circuit Current)

6. A 12-volt inverter rated for 3600W has an 87 percent (.87) efficiency rating. Determine the input current of the inverter when the input voltage has been reduced to 10.5 volts.

Stand-alone inverters produce a constant *ac* output voltage. As the inverter's *dc* input voltage source decreases, current produced by this source increases to maintain a constant ac output voltage.

To determine an inverter's input current rating when an inverter is producing rated power at its lowest input voltage per NEC 690.8(A)(4), the power rating of the inverter is divided by the product of the lowest input voltage and the efficiency rating of the inverter. Using the information given in the question, the input current rating of the inverter is,

$$3600W / 10.5V \times .87 = 394.1A$$

690.8(B)(1) - Sizing of Conductors and Overcurrent Devices (Ampacity and Overcurrent Device Ratings)

7. What size circuit conductors and overcurrent device are required for the system in question No. 3.?

NEC 690.8(B)(1) and (2), respectively requires overcurrent devices and circuit conductors to be sized to carry not less than 125 percent of the maximum current as calculated per NEC 690.8(A)(1).

In question No. 3., once the short-circuit current was calculated, the maximum photovoltaic source circuit current was derived based on an increase of 125 percent. Now NEC 690.8(B)(1) requires the maximum photovoltaic source current to be increased by an additional 125 percent. The Informational Note to NEC 690.8(A) makes mention that when the requirements of NEC 690.8(A)(1) and 690.8(B)(1) are both applied, the resulting multiplication factor is approximately 156 percent which is the product of 125 percent times itself (1.25 x 1.25).

With this being said, the conductors and overcurrent device must be sized based on the following results,

$$675A \times 1.25 \text{ or } 540A \times 1.25 \times 1.25 = 844 \text{ (rounded-up)}$$

Because there were no conditions to apply such as correction factors, NEC 690.8(B)(1)(c) is not applicable and the ampacity of the conductors can be selected without having to compare NEC 690.8(B)(2)(a) and (b).

NEC 690.8(B)(1)(d) also states that the rating or setting of overcurrent devices shall be permitted in accordance with 240.4(B) and 240.4(C). Based on the calculated results (844A), the next standard size overcurrent device per NEC 240.6(A) can be used which would be a 1000A overcurrent device. Based on NEC 240.4(C), the ampacity of the circuit conductors must be either equal to or greater than the rating of the 1000A overcurrent device, that is, if a single 1000A overcurrent device is used. If a combination of overcurrent devices was used instead that is equivalent to the rating of the 1000A overcurrent device (examples – two 500A overcurrent devices, one 600A and one 400A overcurrent device, etc.) the provisions of NEC 240.4(B) can be applied. NEC 240.4(C) only applies when a single overcurrent device exceeds 800A.

690.8(D) - Sizing of Module Interconnection Conductors

This section is mentioned only because it involves an application of 125 percent. An application such as this will possibly require acquiring technical data and pertinent information from the manufacturer of such product to properly size the conductors for module interconnection(s).

690.31(C) - Flexible Cords and Cables and Table 690.31(C) - Correction Factors

8. According to Table 400.5(A) the allowable ampacity for a 14 AWG HSJO Flexible cord is 20A. If a 14/3 cord rated for 105°C was used as circuit conductors for a photovoltaic system and the cord was routed through an area where the ambient temperature reached 132°F, what would be the required ampacity of the cord?

NEC 690.31(C) states that the listed ampacity for flexible cords and cable shall be in accordance with NEC 400.5. However, where such cords and cable are installed in areas where the ambient temperature exceeds 30°C (86°F), the ampacity shall be derated by the correction factors given in Table 690.31(C).

Per Table 690.31(C), conductors rated for 105°C and installed in areas where the ambient temperature is 132°F the correction factor is .77. Therefore, the conductor's derated ampacity is,

$$20A \times .77 = 15.4A$$

690.45(A) - General (Size of Equipment Grounding Conductors)

NEC 690.45(A) requires equipment grounding conductors in photovoltaic source and photovoltaic output circuits to be sized in accordance with Table 250.122. It also states where no overcurrent device is used in the circuit, an assumed overcurrent device rated at the photovoltaic rated short-circuit current shall be used in Table 250.122. The provision of NEC 250.122(B) is not a requirement of NEC 690.45(A).

Refer to Article 250 (Volume 2). See question Nos. 33. and 34.

690.45(B) - Ground-Fault Protection Not Provided (Size of Equipment Grounding Conductors)

9. Refer to question Nos. 35. and 37. of Article 310 (Volume 2). What size equipment grounding conductor would be required if the mentioned conductors where affiliated with a photovoltaic output circuit?

According to NEC 690.45(B) for other than dwelling units where ground-fault protection is not provided in accordance with NEC 690.5(A) - (C), each equipment grounding conductor shall have an ampacity of at least two (2) times the temperature and conduit fill corrected circuit conductor ampacity.

Where the means of ground-fault protection does not exist for photovoltaic systems, short-circuit or ground-fault currents from corresponding source circuits could permit such currents to flow about related equipment grounding conductors up to two times (2 x) the normal circuit current under ground-fault conditions. Refer to the Informational Note following NEC 690.45(B) for additional information.

Therefore, based on the given operating conditions (temperature and conduit fill) of the aluminum conductors and the calculated results of question No. 35., the equipment grounding conductor must have an ampacity that is no less than 218.76A (109.38A x 2).

As it pertains to question No. 37., the selected equipment grounding conductor must have an ampacity that is no less than 273.44A (136.72A x 2).

ARTICLE 692 - Fuel Cell Systems

692.8(B) - Conductor Ampacity and Overcurrent Device Ratings (Circuit Sizing and Current)

The ampacity of the feeder circuit conductors from fuel cell system(s) to premises wiring shall not be less than the greater of (1) nameplate(s) rated current or (2) the rating of the fuel cell system(s) overcurrent protective devices.

692.8(C) - Ampacity of Grounded or Neutral Conductor (Circuit Sizing and Current)

1. While expected to supply an existing load of 22.67A, what size neutral conductor is required to supply a 120V fuel cell output that originates from a 208/120V-1φ system if the fuel cell's output current is rated for 8.5A?

According to NEC 692.8(C), the maximum load connected between the neutral conductor and any one ungrounded conductor plus the fuel cell system output rating must not exceed the ampacity of the neutral conductor if an interactive single-phase, 2-wire fuel cell output is connected to the neutral conductor and one ungrounded conductor of a 3-wire system or of a 3-phase, 4-wire, wye-connected system.

Considering the existing load and the output current of the fuel cell system, the neutral conductor must have a minimum ampacity of 31.17A (22.67A + 8.5A).

692.10(B) - Sizing and Protection (Stand-Alone Systems)

The circuit conductors between the fuel cell system(s) output and the building or structure disconnecting means shall be sized based upon the output rating of the fuel cell system(s). Circuit conductors are to be protected from overcurrents in accordance with NEC 240.4.

ARTICLE 695 - Fire Pumps

695.5(A) - Size (Transformers)

1. A 480-208/120V, 3-phase transformer is needed to supply the fire and jockey pump motors in Figure 695.5(A)-1. Determine the minimum size of the required transformer.

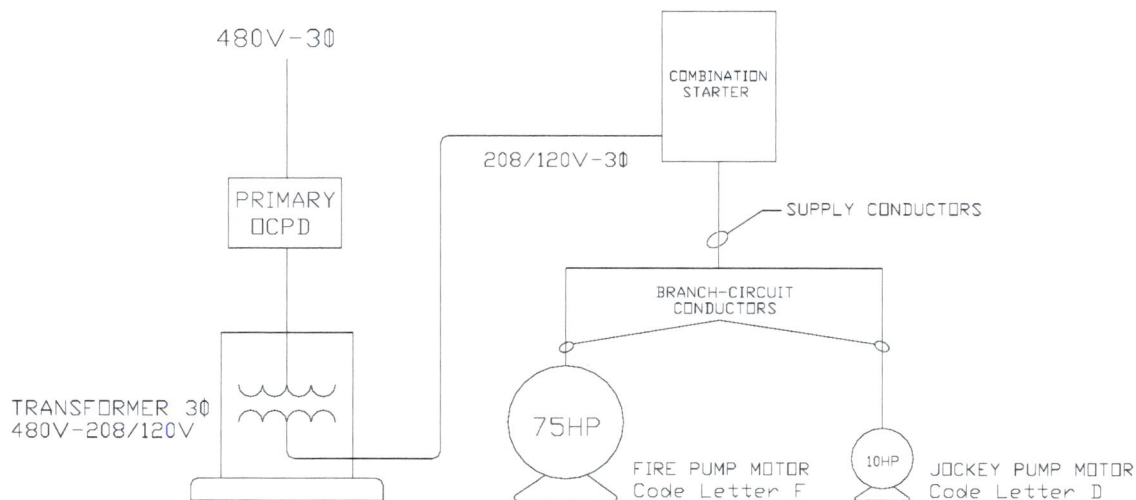

Figure 695.5(A)-1

NEC 695.5(A) requires a transformer supplying an electric motor driven fire pump to be sized at 125 percent of the sum of the fire pump motor and pressure maintenance pump motor loads, and 100 percent of the associated fire pump accessory equipment supplied by the transformer.

To determine the sums of the given motors, Table 430.250 of Article 430 is referenced to gather the full-load current values of both motors.

Per Table 430.250 (3-phase/208V) - 75HP = 211A   10HP = 30.8A

Per NEC 695.5(A)

(211A + 30.8A) x 1.25 = 302.25A.   This calculated value is used to size the required transformer. To determine the size of the required transformer, the following formula is used,

$$kVA = \frac{V \times A \times \sqrt{3}}{1000} \quad \text{where } V = 208V, A = 302.25A \ \sqrt{3} = 1.732$$

$$kVA = \frac{208V \times 302.25A \times 1.732}{1000} = 108.89$$

As a minimum, the transformer must be rated for or exceed 108.89kVA.

695.5(B) - Overcurrent Protection (Transformers)

2. Refer to Figure 695.5(A)-1. What size primary overcurrent device is required to protect the transformer?

As a minimum, the overcurrent device protecting the transformer's primary must be selected *or* set to carry indefinitely the sum of the locked-rotor currents of both motors according to NEC 695.5(B). This is because during the time of a fire the need to protect the fire pump motor is not nearly as severe as for the motor to remain operating to supply water for the cause. See *Exception* to 430.72(C) where control transformers are involved.

In order to determine the size overcurrent device needed, the locked-rotor currents of both motors must be determined per motor code letter and the data given in Table 430.7(B).

75HP motor – Code Letter F   kVA/HP with LR = 5.59 (worst case)
10HP motor – Code Letter D   kVA/HP with LR = 4.49 (worst case)

<u>Using the formula provided in question No. 7 of Article 430,</u> where

$$LRC = \frac{kVA \text{ [listing per Table 430.7(B)]} \times 1000 \times HP}{V \times \sqrt{3}}, \text{ where } \sqrt{3} = 1.732$$

Motors locked-rotor current (LRC)

$$LRC_{(75HP)} = \frac{5.59kVA \times 1000 \times 75HP}{208V \times 1.732} = 1163.76A$$

$$LRC_{(10HP)} = \frac{4.49kVA \times 1000 \times 10HP}{208V \times 1.732} = 124.63A$$

Total LRC of both motors - 1163.76A + 124.63A = 1288.39A

The total LRC for both motors reflects the load side (transformer's secondary) locked-rotor current. Because the overcurrent device is needed for primary protection an equivalent locked-rotor current must be determined based on the primary side.

The primary locked-rotor current can be determined by taking the secondary-to-primary voltage ratio and multiplying this percentage value by the secondary locked-rotor current (1288.39A). Therefore, the primary LRC is determined as follows,

$$\frac{208V}{480V} \times 1288.39A = 558.3A$$

In order to accommodate the calculated results, a standard size 600A overcurrent device per NEC 240.6(A) is required. Because NEC 695.5(B) require the primary protective device to carry indefinitely the sum of the locked-rotor current of the fire pump motor and the pressure maintenance pump motor (558.3A), a 500A overcurrent device could not be used. Even if the primary voltage had been used in the formula instead, the results would have been the same.

695.6(B)(1) - Fire Pump Motors and Other Equipment (Conductor Size)

3. What size conductors are required to supply the fire and jockey pump motors? Consider the use of THHW (75°C) copper conductors.

The requirements for sizing the conductors supplying the fire and jockey pump motors per NEC 695.6(B)(1) are the same as the requirements of NEC 695.5(A) for sizing a transformer. Based on the calculated 302.25A current value for both motors, the size of the supply conductors can be determined. Per Table 310.15(B)(16), as a minimum 350 kcmil THHW copper conductors are required to supply the motors.

695.6(B)(2) - Fire Pump Motors Only (Conductor Size)

4. What size conductors are required, if only the fire pump motor is considered? Consider the use of THHW (75°C) copper conductors.

According to NEC 695.6(B)(2), the conductors are required to be sized in accordance with NEC 430.22. As a result, NEC 430.22(A) requires the conductors to be sized at 125 percent (1.25) of the fire pump motor's full-load current. Therefore, 211A x 1.25 = 263.75A. Per Table 310.15(B)(16), as a minimum 300 kcmil THHW copper conductors are required to supply the fire pump motor alone. NEC 430.22(A) must also be applied to size the branch-circuit conductors for the jockey pump motor.

695.7(A) - Starting (Voltage Drop)

5. Figure 695.5(A)-1. If the fire pump motor experienced a 15 percent voltage drop from its initial source (transformer secondary) to the line terminals of the combination starter (controller) serving the motor, calculate the expected voltage at the line terminals of the controller.

At 208V, a 15 percent (.15) voltage drop would amount to 31.2V (208V x .15). As a result, the voltage at the line terminals of the controller is expected to be 176.8V (208V – 31.2V).

695.7(B) - Running (Voltage Drop)

6. Determine the allowed voltage drop at the fire pump's motor terminals when the motor operates at 115 percent of its full-load current rating.

At 5 percent below the voltage rating of the motor, the voltage at the motor terminals must not drop more than 10.4 volts (208V x .05) when the motor is operating at 115 percent of its full-load current, where 211A x 1.15 = 242.65A. Therefore, at 242.65A, the voltage at the motor terminals must be no less than 197.6 volts (208V – 10.4V).

ARTICLE 725 - Class 1, Class 2, and Class 3 Remote-Control, Signaling, and Power-Limited Circuits

725.41(A)(2) - Other Class 1 Power Sources (Class 1 Circuit Classifications and Power Source Requirements)

1. A Class 1 power source is rated for .5kVA at 28V. What size overcurrent device would be required to protect the power source?

The provisions of NEC 725.41(A)(2) requires a power source other than a transformer be protected by an overcurrent device rated at not more than 167 percent (1.67) of the volt-ampere rating of the source divided by the rated voltage. Therefore,

$$\frac{500VA}{28V} = 17.86V \times 1.67 = 29.83A$$

A 25A device is required. A 30A device would exceed the 167 percent limitation.

725.45(C) - Branch-Circuit Taps (Class 1 Circuit Overcurrent Device Location)

2. A 10 AWG (ZW) Class 1 copper conductor is tapped from the load side of a controlled light and power circuit overcurrent device. Determine the required ampere rating of the overcurrent device to satisfy this installation.

NEC 725.49(A) states that Class 1 conductors larger than 16 AWG shall not supply loads greater than the ampacities given in NEC 310.15. NEC 310.15 reference the Ampacities for Conductors Rated 0-2000 Volts while NEC 310.15(B) reference the Allowable Ampacity Tables, in particular Table 310.15(B)(16).   Table 310.15(B)(16) is the ampacity table applicable to the conductor referenced in question No. 2.

Per Table 310.15(B)(16), a 10 AWG (ZW) copper conductor has a rated ampacity of 35A. According to NEC 725.45(C) where an installation such as in question No. 2. exists, the rating of the overcurrent device must not exceed 300 percent (3) of the ampacity of the Class 1 conductor. Per calculation,

$$35A \times 3 = 105A$$

Based on the calculated results, the overcurrent device must have an ampere rating of 100A or less. An ampere rating exceeding 100A would exceed the 300 percent limitation.

725.45(D) - Primary Side of Transformer (Class 1 Circuit Overcurrent Device Location)

For similar approach see question No. 21. [240.21(C)(1)] of Article 240 (Volume 2).

725.51(A) - Class 1 Circuit Conductors (Number of Conductors in Cable Tray and Raceway, and Ampacity Adjustment)

The ampacity adjustment factors [NEC 310.15(B)(3)(a)] requirements in NEC 725.51(A) is only applicable if Class 1 circuit conductors carry continuous loads in excess of 10 percent (.10) of their rated ampacity.

725.51(B) - Power-Supply Conductors and Class 1 Circuit Conductors (Number of Conductors in Cable Tray and Raceway, and Ampacity Adjustment)

The ampacity adjustment factors [NEC 310.15(B)(3)(a)] requirements in NEC 725.51(B)(1) are only applicable to all conductors where Class 1 circuit conductors carry continuous loads in excess of 10 percent (.10) of the ampacity of each conductor and where the total number of conductors exceeds three (3).

The ampacity adjustment factors [NEC 310.15(B)(3)(a)] requirements in NEC 725.51(B)(2) are only applicable where Class 1 (power-supply) circuit conductors *do not* carry continuous loads in excess of 10 percent (.10) of the ampacity of each conductor and where the number of power-supply conductors exceeds three (3).

CHAPTER 9 - Tables

SIZING RACEWAY AND RACEWAY NIPPLES

HOW RACEWAY IS SIZED

When determining the size of underlined conduit or underlined tubing, regardless of type, the number or combination of conductors being used, all conductors must be counted. Unlike the provisions of NEC 310.15(B)(2), (3) and (5), which only considers certain operating conditions of a conductors, all conductors being installed in raceway must be considered whether current-carrying, non-current-carrying, grounding, bonding, bare or insulated.

NEC 300.17, Informational Note reference applicable NEC sections pertaining to the allowable fill of conductors in raceway. The allowable (percentage) fill determines the number of conductors permitted in a raceway based on the total cross-sectional area of the conductors being installed. The allowable fill must not surpass the allowed cross-sectional area of the raceway. The allowed (percent of) cross-section of conduit and tubing for conductors are given in **Table 1 of Chapter 9**. **Notes to Tables**, Note 1 of Chapter 9 reference **Informative Annex C** for determining the maximum number of conductors of the same size and same type insulation permitted in a particular size conduit or tubing.

Informative Annex C provides 24 separate tables for each type of raceway (conduit or tubing) relative to the type raceway being used to enclose conductors. The 24 tables are equally divided into two sections. One section of the tables is for concentric stranded conductors and the other section is for compact stranded conductors. Compact stranded conductors have a smaller cross-sectional area than concentric stranded conductors of the same wire gauge.

When Informative Annex C does not provide the appropriate raceway/conductor combinations, the maximum number of conductors permitted in conduit or tubing must be determined and calculated based on conductor data provided by the manufacturer.

When the maximum number of conductors permitted in a particular size conduit or tubing is not of the same size and insulation type, the cross-sectional area of each individual conductor must be determined to calculate the minimum trade size of the conduit or tubing being used. **Table 4 of Chapter 9** list the total cross-sectional area and four percent-fill columns ranging from 31 percent to 60 percent for selected conduit and tubing based on trade name and size. The forty percent (40%) fill column of each raceway type is the most frequently used of all columns to determine the minimum trade size of conduit or tubing when 3 or more conductors are to be installed the same raceway. **Table 5** lists the diameters and cross-sectional areas of various types of insulated conductors while **Table 5A** specifically reference compact copper and aluminum building wire.

Where multi-conductor cable, containing two or more conductors is installed in raceway, the cable is treated as one single conductor. For cables that have elliptical cross-sections such as non-metallic sheathed cable (ROMEX) and various types of low voltage cable, the cross-sectional area is calculated based on using the major diameter of the ellipse.

Because all dimensional guidelines discussed in the NEC are minimal, it is often good practice to use raceways that are larger than the minimum size. This further helps to eliminate insulation damage and overcrowding and for future use will provide spare capacity for the installation of additional conductors. Overall, the "larger-than-minimum concept" can be cost effective and reduce labor and material expense.

RACEWAY NIPPLES

Conduit or tubing nipples are defined in **Notes to Tables**, Note 4 of Chapter 9 as having a maximum length of 24". Nipples are installed between boxes, cabinets, meters, and similar enclosures. Conductors installed inside a nipple are permitted to fill up to sixty percent (60%) of the nipple's cross-sectional area. The adjustment factors described in NEC 310.15(B)(3)(a)(2) does not apply to nipples.

conduit - A tube or channel used solely for enclosing and protecting electrical conductors (wires) and cables. May be either rigid or flexible.

tubing - A threadless lightweight form of metal conduit or a flexible circular raceway that has either a metallic or nonmetallic outer surface.

cross-sectional area (csa) - A section of a circular or rectangular object that measures or displays the boundaries of that object as a whole.

concentric stranded conductors - Single, solid wires that have a common center point that form one individual conductor.

compact stranded conductors - Single, solid wire that are closely packed or placed together to form one individual conductor.

> **compact stranding -** The result of a manufacturing process where the stranded conductor is compressed to the extent that the interstices (voids between stranded wires) are virtually eliminated.

multi-conductor cable - An assembly of two or more single conductors that are either individually insulated or bare and covered with a protective jacket for additional protection against chemical, environmental, mechanical and thermal exposure.

> nonmetallic-sheathed cable containing two conductors with an equipment grounding conductor and direct-burial cable are examples of cables having an elliptical shape.

> nonmetallic-sheathed cable containing three conductors with an equipment grounding conductor is an example of a cable having a circular shape. As the wire size of a three conductor nonmetallic-sheathed cable increases, the uniformity of its circular features diminishes.

single conductor - A conductor that either consist of a solid wire or a combination of twisted solid stranded wires that are enclosed in a particular type insulation to form one combined path for the flow of current.

ACCORDING TO THE NEC

The following NEC sections and tables are applicable for understanding how raceway and raceway nipples are sized and selected per Chapter 9 and Informative Annex C.

Table 1 of **Chapter 9** - Percent of Cross Section of Conduit and Tubing for Conductors - Table 1 provides the allowable percentage of raceway fill based on the length of conductor pull and raceway bends that are in reasonable limits. (See Figures CH9-T1A thru CH9-T1C)

> **Table 1** of **Chapter 9, Informational Note No. 2** - Just as all informational notes, Informational Note No. 2 offers sound information to the user. Informational Note No. 2 provides an exclusive procedure for preventing damage to the insulation of conductors by reducing conductor jamming during the pulling of conductors in raceway. Without insulation, an ungrounded conductor is like a car without brakes, totally disastrous.

> **Notes to Tables, Note 3** of Chapter 9 - When equipment grounding or bonding conductors are installed in raceway, the cross-sectional area of these conductors must be included when calculating raceway fill. The cross-sectional area of these conductors whether insulated or bare are provided in Table 4 or Table 8 of Chapter 9 respectively.

> **Notes to Tables, Note 6** of Chapter 9 - When conductors of a different size and insulation type are installed in raceway, Tables 5 and 5A of Chapter 9 must be used to determine the dimensions of the conductors and Table 4 of Chapter 9 must be used to determine the dimensions of raceway.

> **Notes to Tables, Note 7** of Chapter 9 - When calculating the maximum number of conductors of the same size and insulation type that are permitted in raceway and the results of the calculation exceeds a decimal of 0.8 or larger, the next whole number can be used.

Table 2 of **Chapter 9** - Radius of Conduit and Tubing Bends - Table 2 provides a minimum requirement for making uniform bends in conduit and tubing. Section xxx.24 of those articles that reference particular type raceways refer to this table. The purpose of this table is to produce bends that will not damage raceways or cause the internal diameter of raceways to be reduced beyond effective use. Table 2 lists minimum conduit and tubing bending radius' for bends that can be made in one shot (one attempt), multiple shots (requires several bends to make one complete bend, examples: three and four-point saddles) or other type bends that can be manually produced by hand, example: Flexible Metal Conduit (FMC).

Table 4 of **Chapter 9** - Dimensions and Percent Area of Conduit and Tubing - Table 4 provides dimensional data and the cross-sectional area of various type raceway per trade size (metric designator) and allowable fill capacity. (See Figure CH9-T4).

Table 5 of **Chapter 9** - Dimensions of Insulated Conductors and Fixture Wires - Table 5 provides the diameter and cross-sectional area of insulated conductors and fixture wires per size. All conductors listed in Table 5 are the concentric stranded type. (See Figure CH9-T5).

Table 5A of **Chapter 9** - Compact Copper and Aluminum Building Wire Nominal Dimensions and Area - Table 5A provides the diameter and cross-sectional area of compact copper and aluminum wire for selected wire size per bare and insulated type conductors. (See Figure CH9-T5A).

Table 8 of **Chapter 9** - Conductor Properties - Table 8 provides dimensional and technical data pertaining to the diameter, cross-sectional area and resistivity of bare, stranded copper and aluminum conductors. (See Figure CH9-T8).

Informative Annex C - Raceway and Tubing Fill Tables for Conductors and Fixture Wires of the Same Size - Informative Annex C consist of a combination of 24 Tables that list the maximum number of concentric stranded or compact stranded conductors permitted in various types of raceway when the conductors are of the same size and insulation type (See Figure CH9-1). A table of contents is provided in Informative Annex C to reference the page numbers of all tables per raceway types. Table numbers ending with the capital letter "A" represents raceway enclosing compact stranded conductors. [See Figure ANC-TC8/C8(A)].

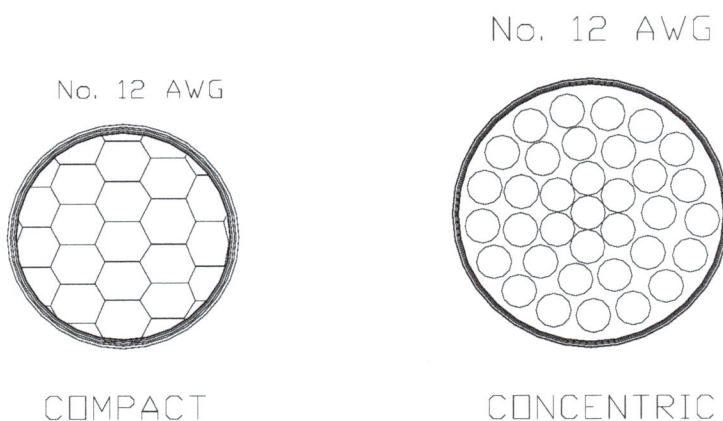

No. 12 AWG

No. 12 AWG

COMPACT

CONCENTRIC

Concentric stranded conductors will have a larger cross-sectional area than a compact stranded conductor having the same AWG size.

Figure CH9-1 (Compact and Concentric stranded conductors)

As applied to Raceway Nipples

NEC 310.15(B)(3)(a)(2) - When conductors are installed in a raceway not exceeding 24" in length ampacity derating is not required.

Notes to Tables, Note 4 of Chapter 9 - When raceway not exceeding 24" is installed between boxes, cabinets and similar enclosures, the raceway is allowed to be filled to 60 percent of its cross-sectional area. (See Figure CH9-N4).

As applied to Multiconductor Cable

Notes to Tables, Note 5 of Chapter 9 - When conductors such as multiconductor and optical fiber cables are not included in Chapter 9 the actual dimensions of the cable must be used.

Notes to Tables, Note 9 of Chapter **9** - When multiconductor cable or flexible cord consist of two or more conductors, the cable must be treated as a <u>single conductor</u> for calculating raceway percent fill capacity (See Figure CH9-2). For cable having elliptical cross-sections such as NM cable (ROMEX), the major diameter of the cable is used as if it was the diameter of a circular cable when calculating cross-sectional area. (See Figures CH9-T1A and CH9-N9).

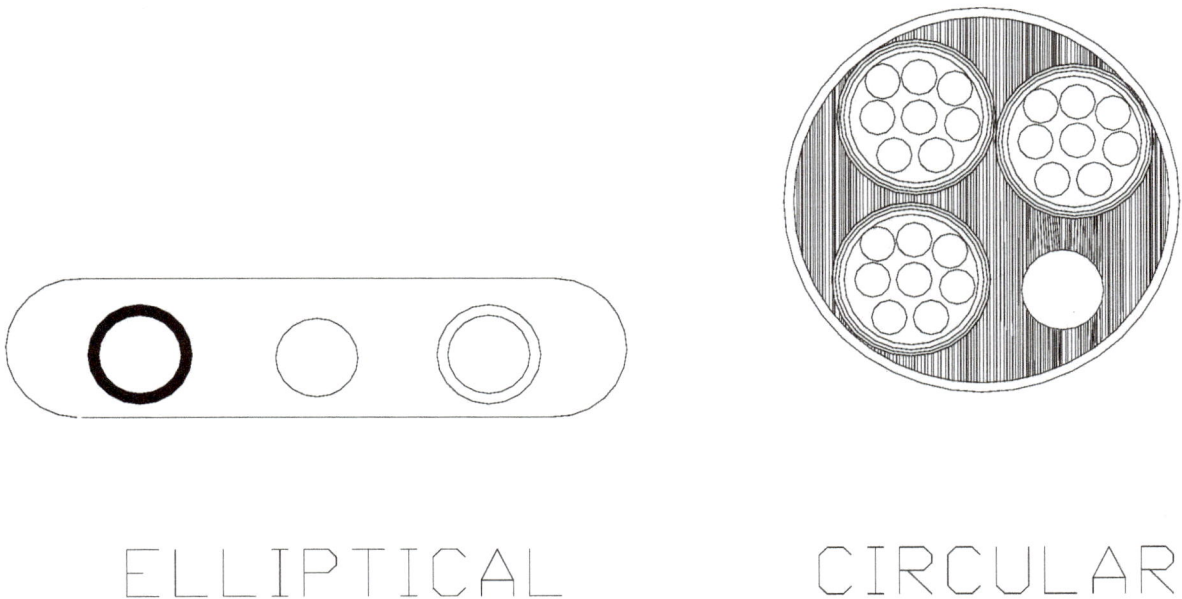

ELLIPTICAL   CIRCULAR

Figure CH9-2 (Multi-conductor cable)

APPLYING THE NEC

SIZING RACEWAY

There are two basic methods for determining raceway fill and sizing raceways.

The *first method* discussed involves conductors of the same size having the same type insulation. The conductors referenced in NEC 310.15(B)(4) and (6) are included in both methods. **Notes to Tables**, Note 3 of Chapter 9 specifically reference the inclusion of equipment grounding or bonding conductors when calculating raceway fill. **Notes to Tables**, Note 1 of Chapter 9 reference the Tables in Informative Annex C for determining the maximum number of conductors and fixtures wires of the same size and insulation type that are permitted in trade sizes of applicable conduit or tubing. The 40 percent (40%) fill capacity as specified in Table 4 of Chapter 9 for raceway enclosing more than 2 conductors has been pre-calculated in each of the Tables in Informative Annex C.

The *second method* involves conductors of a different size and insulation type.

Again, the conductors referenced in NEC 310.15(B)(4) and (6) and **Notes to Tables**, Note 3 of Chapter 9 must also be included when applying the second method.

These methods will be discussed in detail followed by various applications to provide practical knowledge for determining the minimum size raceway needed per application.

All raceway discussed in either method regardless of type or size does not include raceway measured in lengths of 24" or less.

Sizing Raceway for Conductors of Same Size and Insulation Type

When sizing raceway that will enclose the same size conductors having similar type insulation, refer to NEC 300.17, **Notes to Tables**, Note 1 of Chapter 9 and Informative Annex C.

> **NEC 300.17** - Number and Size of Conductor in Raceway - The number and size of conductors in any raceway must not restrict the dissipation of heat nor the installation or withdrawal of conductors without damaging the conductors or its insulation.

Refer to the following procedures for sizing raceway for conductors of same size and insulation type,

1. Determine the number of conductors to be installed in raceway.

Once the number of conductors to be installed in the raceway is determined, **Notes to Tables**, Note 1 of Chapter 9 reference **Informative Annex C** for determining the applicable raceway type per trade size.

2. Refer to **Informative Annex C** and select the applicable table per raceway need and use based on conductor type (concentric stranded or compact stranded), type insulation and conductor size. See Figure ANC-TC.8/C.8(A). Figure ANC-TC.8/C.8(A) serves as an abbreviated guide to explain how the Tables of Informative Annex C are used. An abbreviation of Tables C.8 and C.8(A), Rigid Metallic Conduit (RMC) are used as references.

(A) - Both columns of each table represent the type insulation per letter marking found on the outer surface of the conductor's insulation. With the exception of insulation types RHH, RHW and RHW-2 as identified with an asterisk(*), no other type insulation is enclosed with an outer <u>cover</u>.

> **cover** - The outer material which encloses the insulation of a conductor, yet is not recognized by the National Electrical Code as electrical insulation.

Table C.8 Maximum Number of Fixture Wires in Rigid Metal Conduit (RMC) (Abbreviated)

| | | CONDUCTORS | | | | | | | | | | | |
| | Conductor | Metric Designator (Trade Size) | | | | | | | | | | | |
(A) Type	Size (AWG/ (B) kcmil)	16 (1/2)	21 (3/4)	27 (1)	35 (1-1/4)	41 (1-1/2)	(C) 53 (2)	63 (2-1/2)	78 (3)	91 (3-1/2)	103 (4)	129 (5)	155 (6)
THHN,	12	9	16	26	46	62	102	146	225	301	387	608	877
THWN,	6	2	4	7	12	16	27	38	59	79	101	159	230
THWN-2	2	1	1	3	5	7	11	17	26	34	44	70	100
	3/0	0	1	1	1	3	5	7	11	15	19	30	43
	350	0	0	1	1	1	2	3	5	7	10	15	22
	700	0	0	0	1	1	1	1	3	4	5	8	11
	1000	0	0	0	0	1	1	1	1	3	4	6	8

Table C.8(A) Maximum Number of Compact Conductors in Rigid Metal Conduit (RMC)

Type	Size	16	21	27	35	41	53	63	78	91	103	129	155
THHN,	8	—	—	—	—	—	—	—	—	—	—	—	—
THWN,	6	2	5	8	13	18	30	43	66	88	114	179	258
THWN-2	2	1	1	3	6	8	13	19	29	39	50	79	114
	3/0	0	1	1	2	3	6	8	13	17	22	35	51
	350	0	0	1	1	1	3	4	6	8	11	17	25
	700	0	0	0	1	1	1	1	3	4	6	9	13
	1000	0	0	0	0	1	1	1	2	3	4	6	9

Figure ANC-TC.8/C.8(A) (Tables C.8 and C.8(A) of Informative Annex C)

(B) - These columns list the size of conductors according to American Wire Gauge (AWG) or kilo circular mils (kcmil). In general, the overall AWG conductor sizes for concentric stranded conductors range from 14 to 4/0 and from 8 to 4/0 for compact stranded conductors. Conductors sizes exceeding 4/0 are identified by their cross-sectional area in circular mils beginning with 250,000 cmil (250 kcmil) up to 1,000,000 cmil (1000 kcmil) for compact stranded conductors and up to 2,000,000 cmil (2000 kcmil) for concentric stranded conductors.

(C) - Based on trade sizes (metric designators), each column represents the maximum number of conductors permitted per raceway type.

3. Locate the applicable insulation type per conductor size.

4. Select the appropriate size raceway per number of conductors being installed. (See Figure CH9-3)

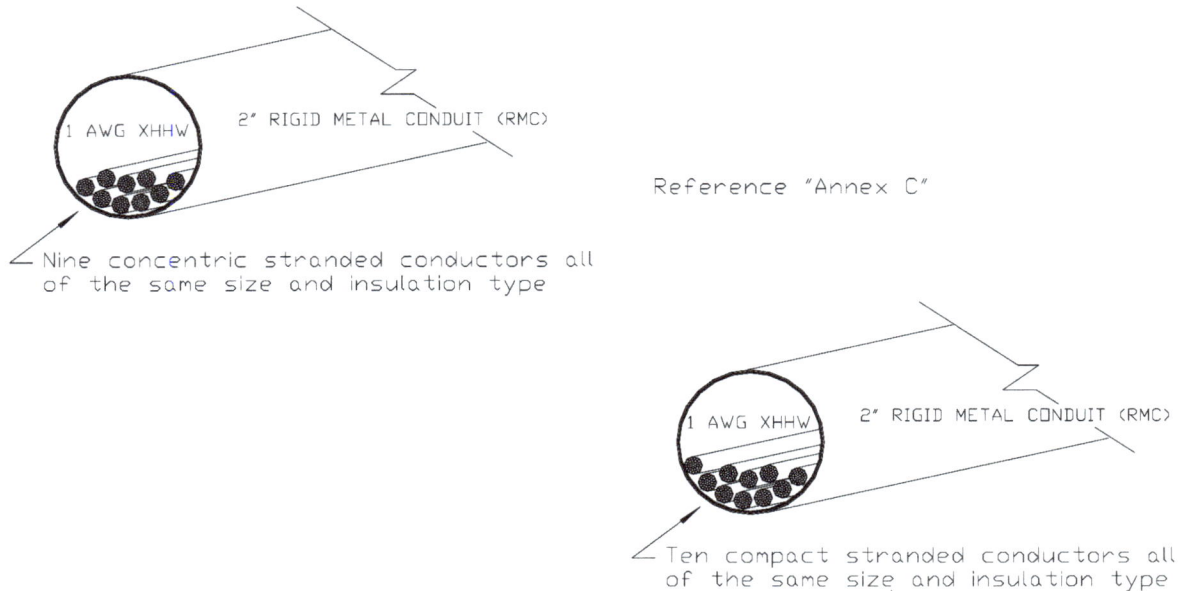

Figure CH9-3 (Rigid Metal Conduit containing maximum number of conductors)

Table 1 - Percent of Cross Section of Conduit and Tubing for Conductors

Table 1 of **Chapter 9,** Informational Note **No. 2**

1. Table C.8 of Informative Annex C permits three 2/0 AWG THWN conductors in trade size 1½" Rigid Metal Conduit (RMC). In accordance with Table 1 of Chapter 9 - Informational Note No. 2; what are the possibilities that the three conductors will jam inside the conduit at some point of being installed?

As described in Table 1 of Chapter 9 - Informational Note No. 2, when pulling three conductors into a raceway, if the ratio of the raceway (inside diameter) to the conductor (outside diameter) is between 2.8 and 3.2, jamming can occur. Table 4 of Chapter 9 list the inside diameter of trade size 1½" Rigid Metal Conduit at 1.624 inches. Table 5 of Chapter 9 list the diameter of 2/0 AWG THWN conductors at .532 inches.

Applying the diameters in ratio form,

$$\frac{1.624 \text{ inches}}{.532 \text{ inches}} = 3.05$$

Being that the results of the ratio falls between 2.8 and 3.2, the possible chances of the conductors jamming inside the raceway at some point of being installed are very high. As a result, the risk of damaging the conductor's insulation is highly possible.

Notes to Tables, Note 1 of **Chapter 9** (Same Size Conductors) (2. - 4.) [3]

2. Refer to Figure 9-2. List the maximum number of conductors allowed in Electrical Metallic Tubing (EMT) per insulation type, conductor size and type stranded conductor per Tables C.1 and C.1(A) of Informative Annex C.

Maximum number per trade size

		¾"		1-1/2"		4"	
Type Letters	**Conductor size**	**C.1**	**C.1(A)**	**C.1**	**C.1(A)**	**C.1**	**C.1(A)**
THW	6	3	3	11	12	81	89
	2	1	1	6	7	44	49
	3/0	1	1	2	3	19	21
	250 kcmil	0	0	1	1	13	14
	700 kcmil	0	0	1	1	5	6
Z	10	11	NA	42	NA	309	NA
	3	2	NA	9	NA	69	NA

Compact conductors enclosed in type Z insulation are not available, therefore are listed as not applicable (NA).

Figure 9-2

3. Four branch circuits consisting of nine 8 AWG THWN copper conductors are being pulled in the same Intermediate Metal Conduit (IMC). Determine the minimum trade size IMC needed.

Question Nos. 3. and 4. will be used as examples to demonstrate the procedure for sizing raceway for conductors of same size and insulation type.

Application

Step 1: Determine the number of conductors to be installed.

The number of conductors to be installed = 9.

Step 2: Refer to Informative Annex C and select the applicable table per raceway need and use based on conductor type, type insulation and conductor size.

Because this situation does not identify the type stranded conductors being used, the use of concentric stranded type conductors is assumed. Table C.4 of Informative Annex C reference Intermediate Metallic Conduit or IMC.

Step 3: Locate the applicable insulation type per conductor size.

Refer to columns 1 and 2 as listed in Table C.4 and locate the type insulation per letter type and conductor size. The conductors being installed have type THWN insulation and are 8 AWG in size.

Step 4: Select the appropriate size raceway per number of conductors being installed.

Refer to the appropriate sub-column of column 3 that reflects the number of conductors equal to or exceeding the number being installed in the raceway. In this case, the minimum trade size IMC permitted to install the nine 8 AWG THWN conductors is 1".

Use 1" IMC to install the nine 8 AWG THWN copper conductors.

4. Four 3/0 AWG XHHW-2 compact conductors are used as the secondary conductors of a transformer. The conductors will be ran from the transformer and terminate in a 225A panelboard. Flexible metal conduit is being used to enclose the conductors. What size flexible metal conduit is required?

Application

Step 1: Determine the number of conductors to be installed.

The number of conductors to be installed = 4.

Step 2: Refer to Informative Annex C and select the applicable table per raceway need and use based on conductor type, type insulation and conductor size.

Table C.3(A) of Informative Annex C reference Flexible Metallic Conduit enclosing compact stranded conductors.

Step 3: Locate the applicable insulation type per conductor size.

Refer to columns 1 and 2 as listed in Table C.3(A) and locate the type insulation per letter type and conductor size. The conductors being installed have insulation type XHHW-2 and are 3/0 AWG in size.

Step 4: Select the appropriate size raceway per number of conductors being installed.

Refer to the appropriate sub-column of column 3 that reflects the number of conductors equal to or exceeding the number being installed in the raceway. In this case, the minimum trade size (metric designator) flexible metal conduit permitted to install the four 3/0 AWG XHHW-2 conductors is 2" based on a maximum allowable number of five 3/0 AWG XHHW-2 conductors.

Use 2" flexible metal conduit to install the four 3/0 AWG XHHW-2 conductors.

Sizing Raceway For Conductors Of Unlike Sizes or Insulation Types

When sizing raceway that will enclose conductors of unlike (different) sizes or type insulation, refer to NEC 300.17, **Notes to Tables**, Notes 3, 5, 6, 7, 8, and 9 of Chapter 9 and Tables 4, 5, 5A and 8 of Chapter 9.

Refer to the following procedures for sizing raceway for conductors of unlike size and insulation type,

1. Refer to the applicable Table(s) (5, 5A, 8) of Chapter 9 and select the approximate cross-sectional area of each unlike conductor.

Figure CH9-T5 serves as an abbreviated guide to explain how **Table 5** is used.

(A) - Alphabetized listing of overall insulation types (bold) per row.

(B) - Insulation types relative to corresponding conductor size.

(C) - Conductor sizes, ranging from 18 AWG to 2000 kcmil.

Table 5 Dimensions of Insulated Conductors and Fixture Wires (Abbreviated)

Type	Size (AWG or kcmil)	Approximate Diameter		Approximate Area	
		mm	(D) in.	mm²	(E) in²
(A) Type:	FFH-2, RFH-1, RFH-2, RHH*, RHW*, RHW-2*, RHH, RHW, RHW-2, SF-1, SF-2, SFF-1, SFF-2, TF, TFF, THHW, THW, THW-2, TW, XF, XFF				
RFH-2, FFH-2	(C) 18 / 16	3.454 / 3.759	0.136 / 0.148	9.355 / 11.10	0.0145 / 0.0172
RHH, RHW, RHW-2	14 / 12	4.902 / 5.385	0.193 / 0.212	18.90 / 22.77	0.0293 / 0.0353
	10 / 3	5.994 / 11.18	0.236 / 0.440	28.19 / 98.13	0.0437 / 0.1521
(B)	1/0 / 2/0	15.80 / 16.97	0.622 / 0.668	196.1 / 226.1	0.3039 / 0.3505
	3/0 / 4/0	18.29 / 19.76	0.720 / 0.778	262.7 / 306.7	0.4072 / 0.4754
	250 / 400 / 600	22.73 / 26.62 / 31.57	0.895 / 1.048 / 1.243	405.9 / 556.6 / 782.9	0.6291 / 0.8626 / 1.2135
	2000	52.53	2.072	2175	3.3719
SF-2, SFF-2	18 / 16	3.073 / 3.378	0.121 / 0.133	7.419 / 8.968	0.0115 / 0.0139

Figure CH9-T5 (Table 5 - Dimensions of Insulated Conductors and Fixture Wires)

(D) - These columns list the approximate outer diameter (measured in millimeters-mm and inches-in.) of each conductor and its surrounding insulation.

(E) - These columns list the approximate cross-sectional area (measured in square millimeters-mm² and square inches-in²) of each conductor and its surrounding insulation.

Figure CH9-T5A serves as an abbreviated guide to explain how Table **T5A** is used.

(A) - Conductor sizes, ranging from 8 AWG to 1000 kcmil.

(B) - These columns list the diameter (measured in millimeters-mm and inches-in.) of bare conductors.

(C) - These columns list the approximate outer diameters and cross-sectional areas (csa) (measured in millimeters-mm and inches-in. and square millimeters-mm^2 and square inches-in^2, respectively) of types THW and THHW insulated compact copper and aluminum building wire.

Table 5A Compact Copper and Aluminum Building Wire Nominal......Areas (Abbreviated)

Size (AWG or kcmil)	Bare Conductor Diameter mm	Bare Conductor Diameter in.	Types THW and THHW Approximate Diameter mm	Types THW and THHW Approximate Diameter in.	Types THW and THHW Approximate Area mm^2	Types THW and THHW Approximate Area in.^{2}	Type THHN Approximate Diameter mm	Type THHN Approximate Diameter in.	Type THHN Approximate Area mm^2	Type THHN Approximate Area in.^{2}	Type XHHW Approximate Diameter mm	Type XHHW Approximate Diameter in.	Type XHHW Approximate Area mm^2	Type XHHW Approximate Area in.^{2}	Size (AWG or kcmil)
8	3.404	0.134	6.477	0.255	32.90	0.0510	–	–	–	–	5.690	0.224	25.42	0.0394	8
4	5.410	0.213	8.509	0.335	56.84	0.0881	7.747	0.305	47.10	0.0730	7.747	0.305	47.10	0.0730	4
1	7.595	0.299	11.81	0.465	109.5	0.1698	10.54	0.415	87.23	0.1352	10.54	0.415	87.23	0.1352	1
1/0	8.534	0.336	12.70	0.500	126.6	0.1963	11.43	0.450	102.6	0.1590	11.43	0.450	102.6	0.1590	1/0
2/0	9.550	0.376	13.84	0.545	150.5	0.2332	12.57	0.495	124.1	0.1924	12.45	0.490	121.6	0.1885	2/0
3/0	10.74	0.423	14.99	0.590	176.3	0.2733	13.72	0.540	147.7	0.2290	13.72	0.540	147.7	0.2290	3/0
250	13.21	0.520	18.42	0.725	266.3	0.4128	17.02	0.670	227.4	0.3525	16.76	0.660	220.7	0.3421	250
300	14.48	0.570	19.69	0.775	304.3	0.4717	18.29	0.720	262.6	0.4071	18.16	0.715	259.0	0.4015	300
350	15.65	0.616	20.83	0.820	340.7	0.5281	19.56	0.770	300.4	0.4656	19.30	0.760	292.6	0.4536	350
400	16.74	0.659	21.97	0.865	379.1	0.5876	20.70	0.815	336.5	0.5216	20.32	0.800	324.3	0.5026	400
500	18.69	0.736	23.88	0.940	447.7	0.6939	22.48	0.885	396.8	0.6151	22.35	0.880	392.4	0.6082	500
750	23.06	0.908	29.21	1.150	670.1	1.0386	27.31	1.075	585.5	0.9076	27.69	1.090	602.0	0.9331	750
1000	26.92	1.060	32.64	1.285	836.6	1.2968	31.88	1.255	798.1	1.2370	31.24	1.230	766.6	1.1882	1000

Figure CH9-T5A (Table 5A - Compact Copper and Aluminum Building Wire Nominal Dimensions and Areas)

(D) - These columns list the approximate outer diameters and cross-sectional areas (csa) (measured in millimeters-mm and inches-in. and square millimeters-mm^2 and square inches-in^2, respectively) of type THHN insulated compact copper and aluminum building wire.

(E) - These columns list the approximate outer diameters and cross-sectional areas (csa) (measured in millimeters-mm and inches-in. and square millimeters-mm^2 and square inches-in^2, respectively) of type XHHW insulated compact copper and aluminum building wire.

Figure CH9-T8 serves as an abbreviated guide to explain how **Table 8** is used. **Table 8** lists the conductor properties for bare and uninsulated stranded conductors.

(A) - **Conductor sizes**, ranging from 18 AWG to 2000 kcmil.

(B) - **Area**, given in circular mils (square millimeters) for AWG sizes 18 to 4/0. Conductors larger than 4/0 are always identified by their cross-sectional area in circular mils opposed to an American Wire Gauge (AWG) size. The circular mils of a conductor are the same regardless of whether solid or stranded.

Conductors

(C) - **Quantity**, (stranding), the quantity 1 represents a bare solid conductor. All other quantities represent the number of individual strands of solid wire twisted together to form a stranded conductor.

(D) - **Diameter**, (stranding), the diameter (measured in millimeters-mm and inches-in.) of each individual strand of solid wire used to form a single stranded conductor.

Table 8 Conductor Properties (Abbreviated)

Size (AWG or kcmil) (A)	Area mm² (B)	Area Circular mils (B)	Quantity (C)	Stranding Diameter mm (D)	Stranding Diameter in. (D)	Overall Diameter mm (E)	Overall Diameter in. (E)	Overall Area mm² (F)	Overall Area in.² (F)	Copper Uncoated ohm/km (G)	Copper Uncoated ohm/kFT (G)	Copper Coated ohm/km (H)	Copper Coated ohm/kFT (H)	Aluminum ohm/km (I)	Aluminum ohm/kFT (I)
18	0.823	1620	1	–	–	1.02	0.040	0.823	0.001	25.5	7.77	26.5	8.08	42.0	12.8
18	0.823	1620	7	0.39	0.015	1.16	0.046	1.06	0.002	26.1	7.95	27.7	8.45	42.8	13.1
12	3.31	6530	1	–	–	2.05	0.081	3.31	0.005	6.34	1.93	6.57	2.01	10.45	3.18
12	3.31	6530	7	0.78	0.030	2.32	0.092	4.25	0.006	6.50	1.98	6.73	2.05	10.69	3.25
6	13.30	26240	7	1.56	0.061	4.67	0.184	17.09	0.027	1.608	0.491	1.671	0.510	2.652	0.808
1	42.41	83690	19	1.69	0.066	8.43	0.332	55.80	0.087	0.505	0.154	0.524	0.160	0.829	0.253
2/0	67.43	133100	19	2.13	0.084	10.62	0.418	88.74	0.137	0.3170	0.0967	0.329	0.101	0.523	0.159
4/0	107.2	211600	19	2.68	0.106	13.41	0.528	141.1	0.219	0.1996	0.0608	0.2050	0.0626	0.328	0.100
500	253	—	37	2.95	0.116	20.65	0.813	336	0.519	0.0845	0.0258	0.0869	0.0265	0.1391	0.0424
1250	633	—	91	2.98	0.117	32.74	1.289	842	1.305	0.0338	0.0103	0.0347	0.0106	0.0554	0.0169
2000	1013	—	127	3.19	0.126	41.45	1.632	1349	2.092	0.02109	0.00643	0.02109	0.00662	0.0348	0.0106

Figure CH9-T8 (Table 8 - Conductor Properties)

(E) - **Diameter**, (overall), the overall diameter (measured in millimeters-mm and inches-in.) of each bare conductor. The overall diameter of a stranded conductor is always slightly larger than that of a solid conductor of the same size.

(F) - **Area**, (overall), the overall (cross-sectional) area (measured in square millimeters-mm² and square inches-in.²) of each bare conductor.

Direct-Current Resistance at 75°C (167°F)

G - **Copper** (Uncoated), the resistance* of a copper conductor per size of conductor without an outer protective coating.

H - **Copper** (Coated), the resistance* of a copper conductor per size of conductor with an outer protective coating. A copper coated conductor has a higher resistance value than an uncoated copper conductor.

I - **Aluminum**, the resistance* of an aluminum conductor per size of conductor. Aluminum conductors have a higher resistance to the flow of current than copper conductors due to the number of free electrons in the outer shell of an aluminum atom compared to a cooper atom.

*The listed resistance values are provided in units of ohms per kilometer-km *or* ohms per kilofeet-kFT. These resistance values are dependent upon the conductors operating in a constant surrounding temperature of 75°C or 167°F and at a length of one thousand feet (1000'). Once the surrounding temperature *or* the length of a conductor changes, the resistance values changes also. See **Notes 1** and **2**.

2. Total the approximate cross-sectional area of conductors.

The cross-sectional area for those conductors of the same size and insulation type must be grouped and totaled separately followed by the total cross-sectional area of all conductors involved. (See Figure CH9-4)

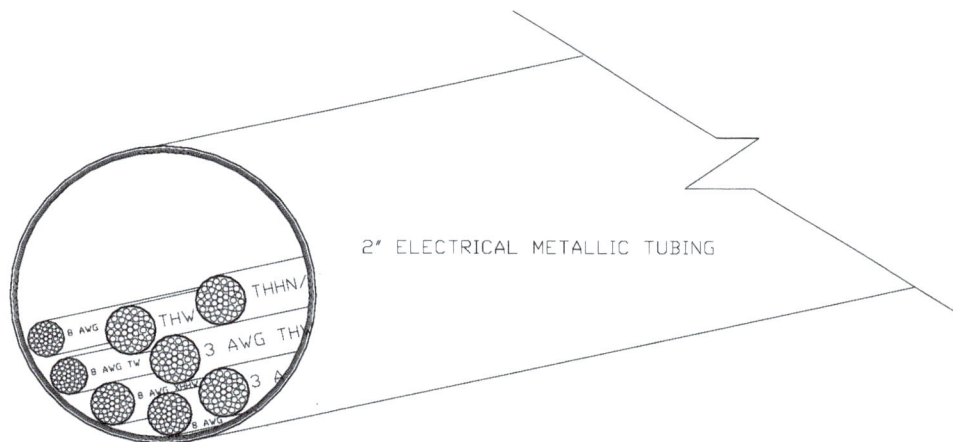

Conductors of different sizes and insulation types

Figure CH9-4

3. Determine the allowable raceway percent fill.

Once the number of conductors to be installed in the raceway has been determined, refer to Table 1 of Chapter 9 for the allowable percent fill per cross-sectional area for conduit or tubing.

Figure CH9-T1 serves as a guide to explain how **Table 1** is used.

Table 1 Percent of Cross Section of Conduit and Tubing for Conductors

Number of Conductors	All Conductor Types
1	53 (A)
2	31 (B)
over 2	40 (C)

Figure CH9-T1 (Table 1 - Percent of Cross Section of Conduit and Tubing for Conductors)

(A) - When a raceway will contain only 1 conductor, the raceway's total cross - sectional area is limited to 53% (.53).

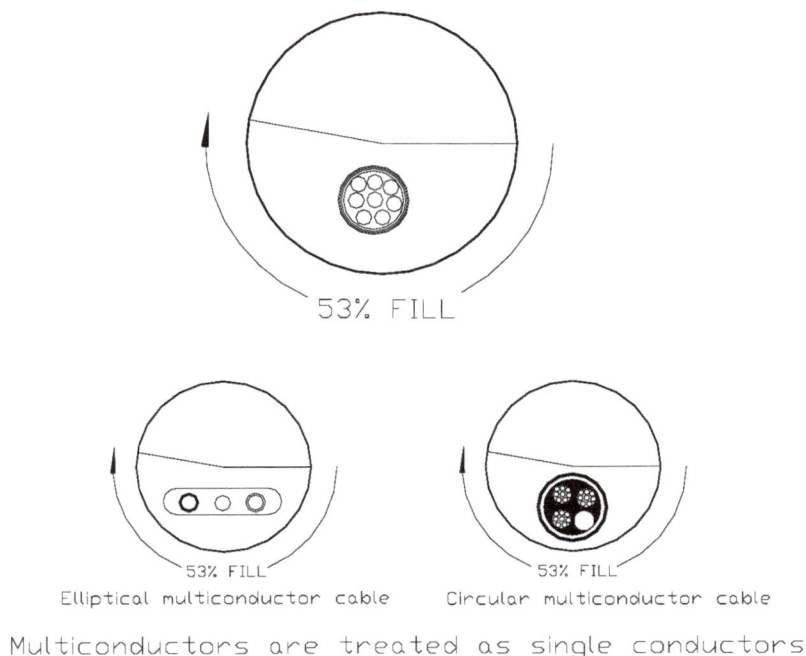

53% FILL

53% FILL
Elliptical multiconductor cable

53% FILL
Circular multiconductor cable

Multiconductors are treated as single conductors

Figure CH9-T1A (Raceway containing only one conductor)

(B) - When a raceway will contain only 2 conductors, the raceway's total cross-sectional area is limited to 31% (.31).

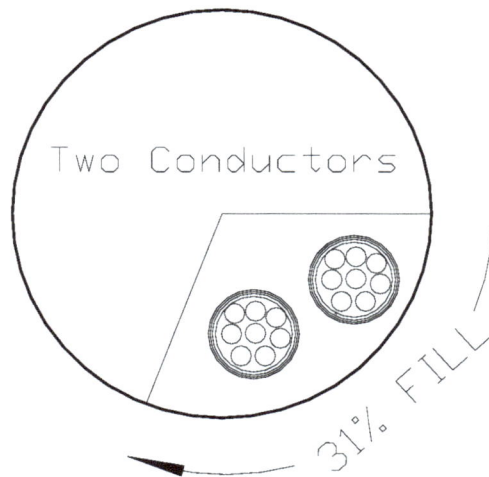

Figure CH9-T1B (Raceway containing only two conductors)

(C) - When a raceway will contain more than 2 conductors, the raceway's total cross-sectional area is limited to 40% (.40).

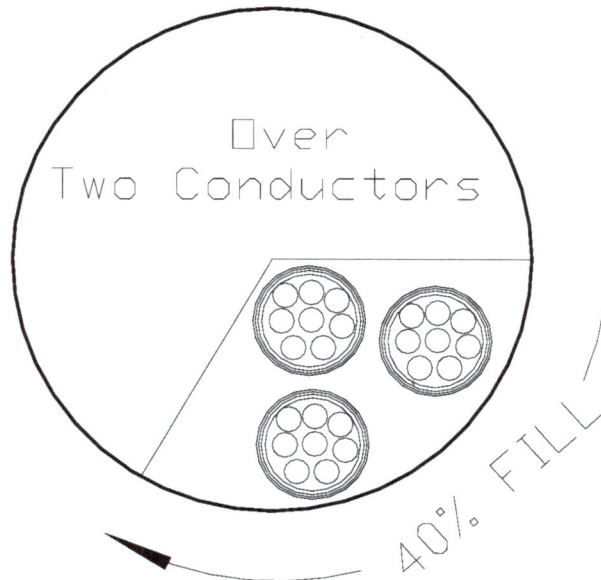

Figure CH9-T1C (Raceway containing more than two conductors)

4. Refer to Table 4 of Chapter 9 (See Figure CH9-T4)

Figure CH9-T4 serves as a guide to explain how **Table 4** is used. Because all inter-raceway tables of Table 4 share the same format, only the raceway table referencing Rigid PVC Conduit (PVC), Schedule 80 (EMT) is used.

(A) - Metric designator. The metric designator is a specific identifier that's equivalent to corresponding raceway trade sizes.

(B) - Listed raceway trade sizes. Sizes vary per raceway type.

(C) - The internal (inside) diameter of each individual raceway per trade size (metric designator) (measured in millimeters-mm and inches-in.).

(D) - The total cross-sectional area (csa) at 100% of each individual raceway per trade size metric designator) (measured in square millimeters-mm^2 and square inches-in^2).

Table 4 Dimensions and Percent Area of Conduit and Tubing
(Areas of Conduit or Tubing for the Combination of Wires Permitted in Table 1, Chapter 9)

Article 352 – Rigid PVC Conduit (PVC), Schedule 80

(A) Metric Designator	(B) Trade Size	(C) Nominal Internal Diameter		(D) Total Area 100%		60%		(F) 1 Wire 53%		(G) 2 Wire 31%		(H) Over 2 Wires 40%	
		mm	in.	mm^2	in.^{2}	mm^2	in.^{2}	mm^2	in.^{2}	mm^2	in.^{2}	mm^2	in.^{2}
12	3/8	—	—	—	—	—	—	—	—	—	—	—	—
16	1/2	13.4	0.526	141	0.217	85	0.130	75	0.115	44	0.067	56	0.087
21	3/4	18.3	0.722	263	0.409	158	0.246	139	0.217	82	0.127	105	0.164
27	1	23.8	0.936	445	0.688	267	0.413	236	0.365	138	0.213	178	0.275
35	1-1/4	31.9	1.255	799	1.237	480	0.742	424	0.656	248	0.383	320	0.495
41	1-1/2	37.5	1.476	1104	1.711	663	1.027	585	0.907	342	0.530	442	0.684
53	2	48.6	1.913	1855	2.874	1113	1.725	983	1.523	575	0.891	742	1.150
63	2-1/2	58.2	2.290	2660	4.119	1596	2.471	1410	2.183	825	1.277	1064	1.647
78	3	72.7	2.864	4151	6.442	2491	3.865	2200	3.414	1287	1.997	1660	2.577
91	3-1/2	84.5	3.326	5608	8.688	3365	5.213	2972	4.605	1738	2.693	2243	3.475
103	4	96.2	3.786	7268	11.258	4361	6.755	3852	5.967	2253	3.490	2907	4.503
129	5	121.1	4.768	11518	17.855	6911	10.713	6105	9.463	3571	5.535	4607	7.142
155	6	145.0	5.709	16513	25.598	9908	15.359	8752	13.567	5119	7.935	6605	10.239

Figure CH9-T4 (Table 4 - Dimensions and Percent Area of Conduit and Tubing.......Chapter 9)

(E) - This column list the allowable csa at 60% of the raceway's total csa, based on a raceway sized for 24" or less (nipple).

(F) - This column list the allowable csa at 53% of the raceway's total csa, based on a raceway containing only 1 wire (conductor).

(G) - This column list the allowable csa at 31% of the raceway's total csa, based on a raceway containing only 2 wires (conductors).

(H) - This column list the allowable csa at 40% of the raceway's total csa, based on a raceway containing more than 2 wires (conductors).

5. Size the raceway per needed type based on the appropriate percent fill column.

The selected value of the appropriate percent fill column should be equal to or greater than the total csa found in step **3**.

For Rewiring or Adding New Wires to Existing Raceway

Refer to Figure CH9-5.

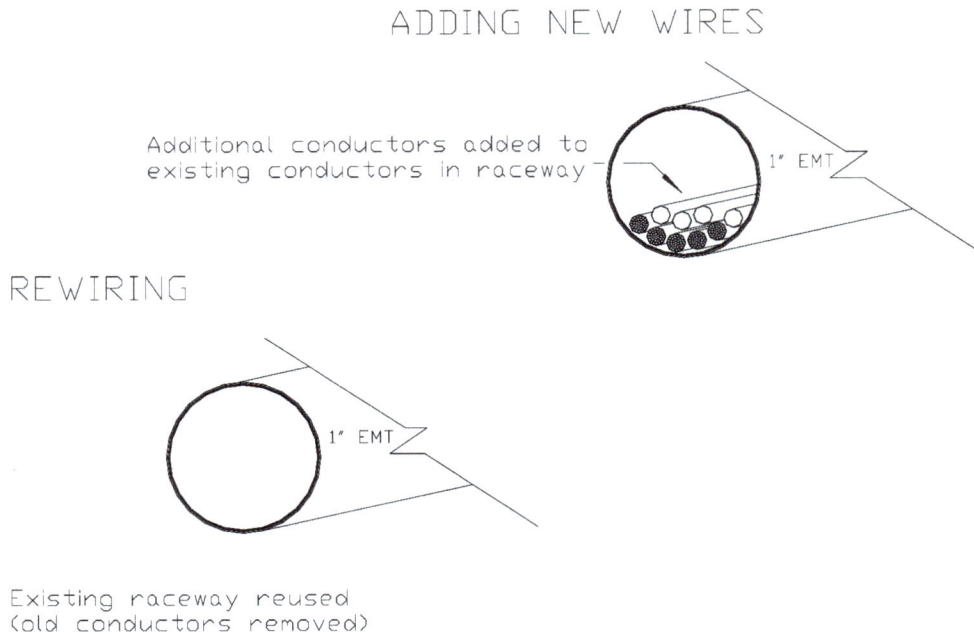

ADDING NEW WIRES

Additional conductors added to
existing conductors in raceway

1" EMT

REWIRING

1" EMT

Existing raceway reused
(old conductors removed)

Figure CH9-5

In the event existing raceway is needed for rewiring or adding new wires and it's unknown whether the existing raceway can be used for such needs, refer to the following procedures,

- Determine the existing raceway's allowable cross-sectional area (csa) based on the required percent fill

- Total the csa of all existing or new rewire conductors to be installed.   Refer to Steps **1. - 3.**, **Sizing Raceway For Conductors of Unlike Size or Insulation Type**.

When rewiring

If the total csa of new rewire conductors is less than the existing raceway's allowable csa, the existing raceway can be re-used, otherwise it has to be replaced with a larger raceway.

or

When adding new wires with existing wires

(a) Subtract the existing conductor's csa from the raceway's allowable csa per required percent fill. This will determine the availability of spare raceway space.

(b) If new conductors to be added are the same (size and insulation type),

    (1) Refer to Table 5 to determine the csa of the new conductors.

    (2) Divide the csa of the new conductor into the available spare raceway space found in **(a)**. This will determine how many additional conductors can be added to the existing installation.

(c) If new conductors to be added are unlike (size and insulation type),

    (1) Total the csa of all existing and new conductors to be added. Refer to Steps **1. - 3.**, **Sizing Raceway For Conductors of Unlike Size or Insulation Type**,

    (2) Compare the total value from the value found in **(a)**. This will determine whether all of the additional conductors can be added to the existing installation,

    (3) If the total csa of all existing and new conductors to be added exceeds the value found in **(a)** the number of new conductors must be reduced until the value is less than that found in **(a)**.

APPLYING THE NEC

SIZING RACEWAY NIPPLES

Notes to Tables, Note 4 of Chapter 9 outlines the maximum length, purpose and percent fill for a raceway nipple. A raceway nipple is similar yet opposite to that of a raceway exceeding 24". Raceway nipples and raceway exceeding 24" both have reduced fill capacities, however conductors enclosed in raceway nipples are not required to be derated according to Notes to Tables, Note 4 of Chapter 9 and to NEC 310.15(B)(3)(a)(2).

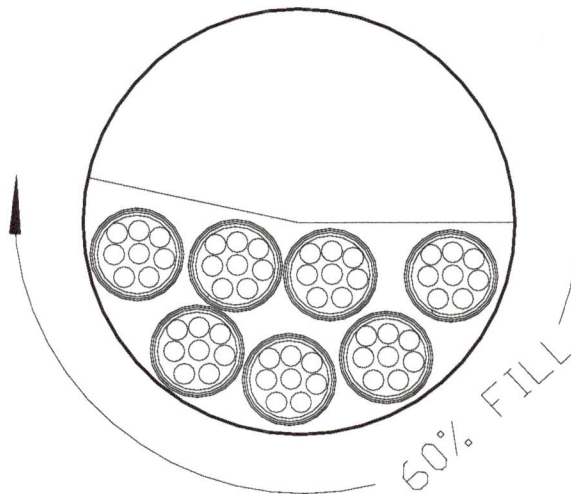

Figure CH9-N4 (Raceway nipple containing multiple conductors)

Prior to the 2002 National Electrical Code, Table 4 of Chapter 9 did not provide a column for listing the cross-sectional percent fill for nipples. This meant that the allowable cross-sectional area of a raceway nipple had to be calculated by using the total cross-sectional area of the raceway at 100 percent (100 %) as listed in each table of Table 4. Knowing that the extra work and the knowledge gained would only serve to enhance one's skills, the decision was made to leave this section in place.

Raceway Nipple Cross-sectional Area

When determining the allowable cross-sectional area of a raceway nipple, refer to **Notes to Tables**, Note 4 of Chapter 9 and Table 4 of Chapter 9.

Refer to the following procedures for determining the allowable cross-sectional area of a raceway nipple,

1. Refer to Table 4.

2. Per raceway table, select the total cross-sectional area (at 100%) of the required raceway per trade size (metric designator).

3. Multiply the value of the total cross-sectional area by 60% (.60).

   The calculated product is the allowable cross-sectional area of a raceway nipple. For conductors being enclosed by a raceway nipple, the conductor's total cross-sectional area must be either equal to or less than the allowable cross-sectional area of the raceway nipple, otherwise a raceway nipple exceeding the total cross-sectional area of the conductors must be used,

   or

4. When applicable, use the following alternative.

   To determine the required total cross-sectional area of a raceway nipple based on the total cross-sectional area of the involved conductors, the following formula is provided:

   $$\text{Total Area}_{(required)} = \frac{\text{Conductor's total cross-sectional area}}{.6}$$

   Once the required total cross-sectional area is determined, refer to the Total Area column of the applicable section per raceway type of Table 4. Select the cross-sectional area in that column that is either equal to or greater than the calculated Total Area value.

Table 2 (Radius of Conduit and Tubing Bends) of **Chapter 9**

5.   Determine the bending radius of a trade size 4 intermediate metal conduit (IMC) and a trade size 3 flexible metal conduit (FMC) if a 90° bend is required to be made with both raceways.

According to NEC 342.24 (Type IMC), the radius of the curve of any field bend to the centerline of the conduit shall not be less than indicated in Table 2 of Chapter 9. Per Table 2, the IMC 90° bend must have a radius of 16 inches. A 90° bend is a one shot (one attempt) bend.

According to NEC 348.24 (Type FMC), the radius of the curve to the centerline of any bend (manually without auxiliary [bending] equipment) shall not be less than shown in Table 2 of Chapter 9 using the column "Other Bends". Per Table 2, referring to the column "Other Bends", the FMC 90° bend must have a radius of 18 inches.

Table 4 (Dimensions and Percent Area of Conduit and Tubing) of **Chapter 9** (6. - 9.) [4]

6.   Determine the internal diameters of the raceways listed below.

Refer to Table 4.

Raceway Type	Trade Size	Internal diameter (in)
IMC	2½	2.557
RMC	4	4.050
RNC, Sch. 80	6	5.709

7.   Refer to Figure 9-7. List the allowable cross-sectional area per raceway percent fill per Table 4.

Trade Size (in.)	EMT	FMC	IMC	LFMC	RMC	RPVC80	.31	.40	.53	.60
3/8		◯					.036	.046	.061	.069
3/8			◯				.059	.077	.102	.115
½	◯						.094	.122	.161	.182
½						◯	.067	.087	.115	.130
1¼				◯			.473	.610	.809	.916
1¼					◯		.510	.659	.873	.988
3		◯					2.191	2.827	3.746	4.241
3½				◯			3.017	3.893	5.158	5.839
4					◯		4.226	5.452	7.224	8.179
5						◯	5.535	7.142	9.463	10.713
6						◯	9.039	11.663	15.454	17.495

Raceway's Allowable Percent Fill (in²)

EMT - Electrical Metallic Tubing   **FMC** - Flexible Metal Conduit
IMC - Intermediate Metal Conduit   **LFMC** - Liquidtight Flexible Metal Conduit
RMC - Rigid Metal Conduit   **RPVC80** - Rigid PVC Conduit, Schedule 80

Figure 9-7

8. Determine the allowable raceway fill of the raceways listed below per Table 4.

Refer to Table 4.

Raceway Type	Trade Size	Allowable Percent Fill	Allowable Raceway Fill (in^2)
EMT	1¼	40%	.598
FMC	1	31%	.253
IMC	4	60%	8.179
LFNC-A	½	53%	.165

9. Calculate the allowable raceway fill of the raceways listed below per total area and csa allowable percent fill. Compare the results with the listed allowable raceway fill values listed in Table 4 per question No. 8.

Refer to Table 4.

Raceway Type	Trade Size	Allowable Percent Fill	Total csa (in^2)	Allowable Raceway Fill (in^2) (All values *rounded off*)
EMT	1¼	40%(.40)	1.496	1.496in^2 x .40 = .598
FMC	1	31%(.31)	.817	.817in^2 x .31 = .253
IMC	4	60%(.60)	13.631	13.631in^2 x .60 = 8.179
LFNC-A	½	53%(.53)	.312	.312in^2 x .53 = .165

Tables 5 (Dimensions of Insulated Conductors and Fixture Wires) *and* **5A** (Compact Copper and Aluminum Building Wire Nominal Dimensions and Areas) of **Chapter 9**

10. Refer to Figure 9-10. List the approximate cross-sectional area of the given conductors per Tables 5 and 5A.

Type Letters	Copper/ Aluminum (Table 5)	Compact Copper/Aluminum (Table 5A)	Conductor Size	Conductor's CSA (in^2)
FFH-2	○		16	.0172
THW	○		8	.0437
THW		○	8	.0510
RHW	○		3/0	.4072
RHW*	○		3/0	.3117
THHN		○	6	.0452
XF	○		14	.0139
THWN-2	○		500 kcmil	.7073
XHH	○		1	.1534

*without outer covering

Figure 9-10

Notes to Tables, Note 4 (Conduit or Tubing Nipples) *and* **Table 4** of **Chapter 9**

11. Refer to Figure 9-11. Calculate each nipples maximum allowable cross-sectional area per trade size and type. Compare the results with those of Table 4.

Application

Determine each raceway nipple's allowable cross-sectional area based on percent fill and NEC references.

Trade Size (in.)	Raceway Type			Total (100%) CSA (in²)	Multiplier .60	Nipple's Allowable CSA (in²) (All values *rounded off*)
	EMT	IMC	RMC			
½		◯		.342	x .60	.205
¾			◯	.549	x .60	.329
1	◯			.864	x .60	.518
1¼			◯	1.526	x .60	.916
1½		◯		2.225	x .60	1.335
2	◯			3.356	x .60	2.014

Figure 9-11

Notes to Tables, Note 4 (Conduit or Tubing Nipples) *and* **Tables 4** and **5** of **Chapter 9**
(12. - 14.) [3]

Question Nos. 12. - 15. and 18. - 23. will be used as examples to demonstrate the procedure for sizing raceway for conductors of unlike size and insulation type.

12. An electrician desires to use a 1" rigid steel conduit nipple to mount between a meter can and a gutter. The conductors that will be enclosed within the nipple have a total cross-sectional area of .6328 in². Is a 1" nipple large enough? If not what size nipple should be used?

Application

Step 1: Refer to Table 4.

Locate the applicable table that will provide the needed data for Rigid Steel Conduit.

Step 2: Per raceway table, select the total cross-sectional area (at 100%) of the required raceway per trade size (metric designator).

The total cross-sectional area of a 1" rigid steel conduit is .887 in².

Step 3: Multiply the value of the total cross-sectional area by 60% (.60).

$$.887 \text{ in}^2 \text{ x } .60 = .5322 \text{ in}^2$$

.5322 in^2 is the allowable cross-sectional area of the raceway nipple per calculation and table listing.

Compared to the total cross-sectional area of the conductors (.6328 in^2), the allowable cross-sectional area (.5322 in^2) of the 1" rigid steel conduit nipple is not large enough for this installation. Therefore, as a minimum, the next size raceway must be used that exceeds the conductor's total cross-sectional area,

or

to determine the correct size nipple that should be used for this installation, use the alternative given in **Step 4**, **Raceway Nipple Cross-sectional Area** where,

$$\text{Total Area}_{(required)} = \frac{.6328 \text{ in}^2}{.6} = 1.0547 \text{ in}^2$$

Refer to the Total Area column of the Rigid Metal Conduit table of **Table 4** and select the cross-sectional area that is either equal to or greater than the calculated Total Area value.

The cross-sectional area listed in the Total Area column of the Rigid Metal Conduit table that is either equal to or greater than the Total Area calculated is 1.526 in^2.

Based on 1.526 in^2 a 1¼" nipple should be used instead.

13. How many 350 kcmil THWN conductors can be enclosed in a 3" EMT nipple?

Application (modified procedures)

Step 1: Refer to Table 4.

Locate the applicable section which will provide the needed data for Electrical Metallic Tubing (EMT).

Step 2: Per raceway table, select the total cross-sectional area (at 100%) of the required raceway per trade size (metric designator).

The allowable cross-sectional area of 3" EMT at 60% is 5.307 in^2.

Step 3: Determine the cross-sectional area of a 350 kcmil THWN conductor.

Per Table 5 of Chapter 9 the cross-sectional area of a 350 kcmil THWN conductor is .5242 in^2.

Step 4: Divide the cross-sectional area of the conductor into the nipples allowable cross-sectional area to determine the number of conductors.

$$5.307 \text{ in}^2 / .5242 \text{ in}^2 = 10.124$$

Where rounded-down, ten (10) 350 kcmil THWN conductors can be enclosed in a 3" EMT nipple.

14. An IMC nipple contains 6-4/0 AWG, 4-250 kcmil and 3-300 kcmil conductors. Determine the required cross-sectional area of the IMC nipple. All conductors have XHHW insulation.

Application (modified procedures)

Determine the total cross-sectional area of the conductors per NEC reference.

Type Letters	Conductor Size	Conductor's CSA (in^2)	No. of conductors per type	Total CSA (in^2) per type
			X	=
XHHW	4/0 AWG	.3197	6	1.9182
XHHW	250 kcmil	.3904	4	1.5616
XHHW	300 kcmil	.4536	3	1.3608

Combined total = 4.8406

To determine the correct size nipple that should be used for this installation, use the alternative given in **Step 4**, **Raceway Nipple Cross-sectional Area** where,

$$\text{Total Area}_{(required)} = \frac{4.8406 \text{ in}^2}{.6} = 8.0677 \text{ in}^2$$

Refer to the Total Area column of the Intermediate Metal Conduit (IMC) table of Table 4 and select the cross-sectional area that is either equal to or greater than the calculated Total Area value.

The cross-sectional area listed in the Total Area column of the Intermediate Metal Conduit table that is either equal to or greater than the Total Area calculated is 10.584 in^2.

Based on 10.584 in^2, a 3½" (91) IMC nipple should be used at minimum.

Notes to Tables, Note 6 (Combination of Conductors) *and* **Tables 4, 5** and **5A** of **Chapter 9**

15. Determine the minimum size raceway required to enclose 4-14 AWG RHH copper (without outer covering) conductors, 5-12 AWG ZW copper conductors and 4-2 AWG THHN compact aluminum conductors.

Application

Step 1: Refer to the applicable Tables (5 and 5A) of Chapter 9 and select the approximate csa of each unlike conductor.

Type Letters	Conductor Size	Conductor's csa (in^2)
RHH	14 AWG	.0209
ZW	12 AWG	.0181
THHN	2 AWG	.1017

Step 2: Total the approximate csa of conductors.

Multiply each conductor's csa per type *times* the number of conductors to produce a total per type. Once this is completed add the totals together to derive conductor's combined csa.

Type Letters	Conductor Size	Conductor's csa (in^2)		Total No. of conductors per type	csa (in^2) per type
			x		=
RHH	14 AWG	.0209		4	.0836
ZW	12 AWG	.0181		5	.0905
THHN	2 AWG	1017		4	.4068
				Combined total =	.5809

Step 3: Determine the allowable raceway percent fill per Table 1 of Chapter 9.

According to Table 1 of Chapter 9 when a raceway will contain over 2 conductors the percent of cross section must be 40%.

Step 4: Refer to Table 4 of Chapter 9.

Because the question did not identify a particular type raceway, reference the first csa value listed under the 40% column that is either equal to or greater than the combined csa value (.5809 in^2) in each table of Table 4.

Step 5: Size the raceway per needed type based on the appropriate percent fill column.

The first csa value that is either equal to or greater than the combined csa value (.5809 in^2) under the 40% column of each table is:

Raceway Type	csa @ 40% fill (in^2)
Electrical Metallic Tubing	.598
Electrical Nonmetallic Tubing	.774
Flexible Metal Conduit	.743
Intermediate Metal Conduit	.659
Liquidtight Flexible Nonmetallic Conduit (Type LFNC-B)	.611
Liquidtight Flexible Nonmetallic Conduit (Type LFNC-A)	.601
Liquidtight Flexible Metal Conduit	.611
Rigid Metal Conduit	.610
Rigid PVC Conduit, Schedule 80	.684
Rigid PVC Conduit, Schedule 40	.581
Type A, **Rigid PVC Conduit**	.707
Type EB, **PVC Conduit**	1.550

Based on the csa values listed above the minimum size raceways are:

Raceway Type	Minimum Trade Size
Electrical Metallic Tubing	1¼ "
Electrical Nonmetallic Tubing	1½"
Flexible Metal Conduit	1½"
Intermediate Metal Conduit	1¼ "
Liquidtight Flexible Nonmetallic Conduit (Type FNMC-B)	1¼ "
Liquidtight Flexible Nonmetallic Conduit (Type FNMC-A)	1¼ "
Liquidtight Flexible Metal Conduit	1¼ "
Rigid Metal Conduit	1¼ "
Rigid PVC Conduit, Schedule 80	1½"
Rigid PVC Conduit, Schedule 40	1¼ "
Type A, **Rigid PVC Conduit**	1¼ "
Type EB, **PVC Conduit**	2"

Notes to Tables, Note 7 of Chapter 9 (Same Size Conductors Calculation Results to 0.8 or Larger Decimal)

16. How many 8 AWG XHHW conductors are allowed to be installed in ½" Rigid Metal Conduit (RMC)? Refer to Tables 4 and 5 of Chapter 9.

$$\frac{\frac{1}{2}" \text{ RMC cross-sectional area @ 40\% (Table 4) - } .125\text{in}^2}{8 \text{ AWG XHHW cross-sectional area (Table 5) - } .0437\text{in}^2} = 2.86 = 3$$

Notes to Tables, Note 7 of Chapter 9 was primarily adopted to justify the whole number values given in Informative Annex C. See Table C.6 of Informative Annex C.

According to **Notes to Tables, Note 8** of **Chapter 9** (Bare Conductors)

17. Determine the diameters and cross-sectional areas of the bare conductors listed below.

According to Notes to Tables, Note 8 of Chapter 9, the dimensions (measurements) for bare conductors in Table 8 are permitted.

Conductor Size	Diameter (in)	Cross-sectional area (in^2)
12 AWG (solid)	.081	.005
8 AWG (stranded)	.146	.017
1 AWG	.332	.087
3/0 AWG	.470	.173
350 kcmil	.681	.364
1000 kcmil	1.152	1.042

Notes to Tables, **Note 3** (Equipment ground or bonding Conductors), **Note 6** (Combination of Conductors) *and* **Note 8** (Bare Conductors) *and* **Tables 4, 5** and **8** of **Chapter 9** (18. - 19.) [2]

18. Determine the minimum size Rigid PVC Conduit (Schedule 80) needed to install 8-3/0 AWG XHHW conductors, 6-6 AWG THW conductors, 4-10 AWG RHW conductors and 1-8 AWG bare copper (solid) equipment grounding conductor.

Application

Step 1: Refer to the applicable Tables (5, 5A and 8) of Chapter 9 and select the approximate csa of each unlike conductor.

Type Letters	Conductor Size	Conductor's csa (in^2)
XHHW	3/0 AWG	.2642
THW	6 AWG	.0726
RHW	10 AWG	.0437
Bare	8 AWG	.0130

Step 2: Total the approximate csa of conductors.

Multiply each conductor's csa per type *times* the number of conductors to produce a total per type. Once this is completed add the totals together to derive the conductor's combined csa.

Type Letters	Conductor Size	Conductor's csa (in^2)		No. of conductors per type		Total csa (in^2) per type
			x		=	
XHHW	3/0 AWG	.2642		8		2.1136
THW	6 AWG	.0726		6		.4356
RHW	10 AWG	.0437		4		.1748
Bare	8 AWG	.0130		1		.0130
					Combined total =	2.7370

Step 3: Determine the allowable raceway percent fill and refer to Table 4 of Chapter 9.

According to Table 1 of Chapter 9 when a raceway will contain over 2 conductors the percent of cross section must be 40%.

Step 4: Refer to Table 4 of Chapter 9.

Locate the applicable table that will provide the needed data for Rigid PVC Conduit, Schedule 80.

Step 5: Size the raceway per needed type based on the appropriate percent fill column.

Select the first csa value listed under the 40% column that is equal to or greater than the combined csa value, 2.7370 in^2. The first csa value that is either equal to or greater than the combined csa value under the 40% column is 3.475 in^2. Based on 3.475 in^2, a 3½" raceway is referenced. Use 3½" Schedule 80-Rigid PVC Conduit to install the conductors.

19. The conductors installed in the 3½" PVC conduit in question No. 18. were removed due to a design change. Can the conduit be re-used to install the following conductors, 12-2/0 AWG THWN conductors, 5-4 AWG XHHW conductors, 4-8 AWG THW conductors and 6-12 AWG RHW conductors?

Application

Refer to the procedures for **Rewiring** or **Adding New Wires** to **Existing Raceway** and proceed.

- Determine the existing raceway's allowable csa based on the required percent fill.

  Referring back to question No. 18., the existing 3½" raceway has a csa of 3.475 in^2.

- Total the csa of all new rewire conductors to be installed as described in steps **1. - 3.**, **Sizing Raceway For Conductors of Unlike Size or Insulation Type**.

  Multiply each conductor's csa per type *times* the number of conductors to produce a total per type. Once this is completed add the totals together to produce the conductor's combined csa.

Type Letters	Conductor Size	Conductor's csa (in^2)	No. of conductors per type	Total csa (in^2) per type
			x =	
THWN	2/0 AWG	.2223	12	2.6676
XHHW	4 AWG	.0814	5	.4070
THW	8 AWG	.0437	4	.1748
RHW	12 AWG	.0353	6	.2118
			Combined total =	3.4612

Refer to procedures "**When rewiring**".

If the total csa of new rewire conductors is less than the existing raceway's allowable csa, the existing raceway can be re-used; otherwise it has to be replaced with a larger raceway.

$$
\begin{array}{lll}
\text{Existing raceway's csa} & = & 3.4750 \text{ in}^2 \\
\text{New rewire conductor's csa} = & & \underline{-3.4612 \text{ in}^2} \\
\text{Difference} & & .0138 \text{ in}^2
\end{array}
$$

Because the total cross-sectional area (3.4612 in^2) of the new rewire conductors is less than the existing raceway's allowable cross-sectional area (3.475 in^2) the 3½" Schedule 80-Rigid PVC conduit can be re-used.

According to **Notes to Tables**, **Note 9** *and* **Table 4** of **Chapter 9** (20. - 21.) [2]

20. Determine the minimum size electrical metallic tubing needed for each multiconductor cable. The diameter of a 10/3 Romex (nonmetallic-sheathed) cable with an equipment grounding conductor is 9/16"(.5625"), whereas the largest diameter of a 10/2 Romex with an equipment grounding conductor is 7/16"(.4375") and the smallest diameter 3/16"(.1875").

Application (modified application)

Step 1: Refer to **Notes to Tables**, Note 9 of Chapter 9.

When multiconductor (Romex) cable contains three conductors with an equipment ground, the shape of the cable is approximately circular. For direct burial and two conductor cable with ground, the shape of the cable is elliptical, see Figure CH9-N9.

According to Note 9 a multiconductor cable consisting of two or more conductors must be treated as a single conductor when calculating the allowable percent fill capacity for conduit and tubing. For multiconductor cable having an elliptical cross-section, the cross-sectional area must be based on the major diameter of the ellipse. Unlike the 10/3 multiconductor cable which has only one diameter to consider, the largest of the two given diameters for the 10/2 multiconductor cable must be used to calculate the cross-sectional area of the cable.

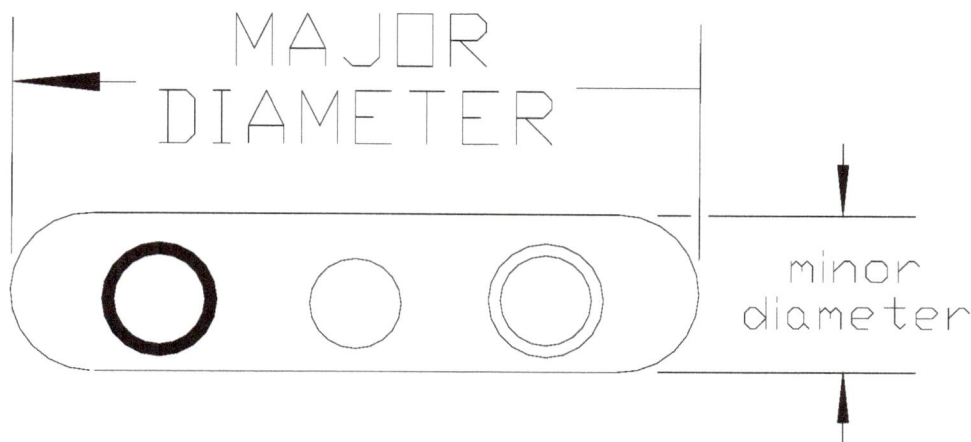

MAJOR AND MINOR DIAMETERS
OF ELLIPTICAL CABLE

Figure CH9-N9

Regardless of shape, the cross-sectional area of both cables can be determined using the following formula,

$$\textbf{Area}_{(circle)} = \Pi\ r^2$$

where, Π (pie) = 3.1416 and **r** = diameter / 2

Therefore, the area (cross-sectional) of the 10/2 cable is,

$$3.1416 \times (.4375"/2)^2 = .1503\ in^2\ (major\ diameter)$$

and the area (cross-sectional) of the 10/3 cable is,

$$3.1416 \times (.5625"/2)^2 = .2485\ in^2$$

Step 2: Determine the allowable raceway percent fill.

According to Table 1 of Chapter 9 when a raceway will contain only 1 conductor the percent of cross section must be 53% (.53).

Step 3: Refer to Table 4 of Chapter 9.

Locate the applicable table that will provide the needed data for Electrical Metallic Tubing.

Step 4: Size the raceway per needed type based on the appropriate percent fill column.

Select the first csa value listed under the 53% column that is greater than the calculated csa values, .1503 in^2 and .2485 in^2. The first csa values greater than the calculated values under the 53% column are .161 in^2 and .283 in^2.

Based on .1503 in^2 and .283 in^2, a ½" and ¾" raceway should be used respectively. Use ½" EMT for the 10/2 cable and ¾" EMT for the 10/3 cable.

21. A 10/4 SJTO flexible cord has an overall diameter of .652 inches. What size intermediate metal conduit is needed to enclose the cord?

Application (modified procedures)

Step 1: Refer to **Notes to Tables**, Note 9 of Chapter 9.

According to Note 9, a flexible cord consisting of two or more conductors must be treated as a single conductor when calculating the allowable per cent fill capacity for conduit and tubing.

Just as in question No. 20., the cross-sectional area of this flexible cord can also be determined using the following formula,

$$\textbf{Area}_{(circle)} = \Pi \; \textbf{r}^2$$
where, Π (pie) = 3.1416 and **r** = diameter / 2

Therefore, the area (cross-sectional) of the 10/4 flexible cord is,

$$3.1416 \times (.652"/2)^2 = .334 \; in^2$$

Step 2: Determine the allowable raceway percent fill.

According to Table 1 of Chapter 9 when a raceway will contain only 1 conductor the percent of cross section must be 53% (.53).

Step 3: Refer to Table 4 of Chapter 9.

Locate the applicable table that will provide the needed data for Intermediate Metal Conduit.

Step 4: Size the raceway per needed type based on the appropriate percent fill column.

Select the first csa value listed under the 53% column that is greater than the calculated csa value, .334 in^2. The first csa value greater than the calculated value under the 53% column is .508 in^2. Based on .508 in^2, a 1" intermediate metal conduit is required to enclose the cord.

Notes to Tables, **Note 6** *and* **Table 4** of **Chapter 9**

22. A 1¼" rigid steel conduit contains 6-14 AWG RHW (with outer covering), 5-12 AWG TW and 3-10 AWG THHW conductors. As a result, how many 8 AWG RHW conductors can be added in the conduit?

Application

Refer to the procedures for **Rewiring** or **Adding New Wires** to **Existing Raceway** and proceed.

- Determine the existing raceway's allowable csa based on the required percent fill.

  Referring to Table 4 of Chapter 9, the allowable csa of a 1¼" Rigid Metal Conduit based on 40% fill is .610 in^2.

- Total the csa of all existing conductors as described in steps **1. - 3.**, **Sizing Raceway For Conductors of Unlike Size or Insulation Type**.

  Multiply each conductor's csa per type *times* the number of conductors to produce a total per type. Once this is completed add the totals together to produce the conductor's combined csa.

Type Letters	Conductor Size	Conductor's csa (in^2)		No. of conductors per type	Total csa (in^2) per type
			x		=
RHW	14 AWG	.0293		6	.1758
TW	12 AWG	.0181		5	.0905
THHW	10 AWG	.0243		3	.0729
				Combined total =	.3392

Refer to procedures "**When adding new wires with existing wires**".

   (a) Subtract the existing conductor's csa from the raceway's allowable csa to determine the availability of spare raceway space.

$$
\begin{array}{lcl}
\text{Existing raceway's csa} & = & .6100 \text{ in}^2 \\
\text{Added conductor's csa} & = & -.3392 \text{ in}^2 \\
\text{Available space} & & .2708 \text{ in}^2
\end{array}
$$

   (b) If new conductors to be added are the same (size and insulation type)

   (1) Refer to Table 5 to determine the csa of the new (added) conductors.

   The cross-sectional area of a 8 AWG RHW conductor is .0835 in^2.

(2) Divide the csa of the new conductor into the available spare raceway space to determine how many additional conductors can be added to the existing installation.

$$.2708 \text{ in}^2 / .0835 \text{ in}^2 = 3.243$$

Because 3.243 is not allowed to be rounded up, a total of three 8 AWG RHW conductors can be added to the existing 1¼" rigid steel conduit. If the number was rounded up to 4 it would cause the 1¼" rigid steel conduit's available space to be exceeded, therefore exceeding the conduit's 40% fill capacity.

Notes to Tables, **Note 6** *and* **Tables 4** and **5** of **Chapter 9**

23. Can 12-14 AWG and 5-10 AWG THWN conductors be added to the 1¼" rigid steel conduit in question No. 22. if the 8 AWG RHW conductors are removed?

Application

Refer to procedures "**When adding new wires with existing wires**".

Because the availability of spare raceway space for the 1¼" rigid steel conduit is known and the new added conductors are unlike; this process can began at step **(c)** of the referenced procedures.

(c) If new conductors to be added are unlike (size and insulation types),

(1) Total the csa of all existing and new conductors to be added. Refer to **steps 1. - 3.**, **Sizing Raceway For Conductors of Unlike Size or Insulation Type**.

Existing conductor's csa = .3392 in² (per question No. 22.)

New conductor's csa

Type Letters	Conductor Size	Conductor's csa (in²)		No. of conductors per type	csa (in²) per type
			x		=
THWN	14 AWG	.0097		12	.1164
THWN	10 AWG	.0211		5	.1055
				Combined total =	.2219

(2) Compare the total value from the value found in **(a)** of question No. 22. This will determine whether all of the additional conductors can be added to the existing installation.

(a) Available space = .2708 in²
New conductor's csa = .2219 in²

(3) If the total csa of all existing and new conductors to be added exceeds the value found in **(a)** the number of new conductors must be reduced until the value is less than that found in **(a)**.

Because the csa of the new conductors is less than the available spare raceway space, the new conductors can be added to the existing 1¼" rigid steel conduit leaving .0489 in² of available spare space remaining (.2708 in² - .2219 in²).

Table 8 (Conductor Properties) of **Chapter 9** (24. - 28.) [5]

24. According to Table 8, a 10 AWG solid conductor has a 0.102 inch diameter. Considering the given dimension, observe how the conductor's circular mils are derived.

The conductor's circular mils is first determined by multiplying the diameter of the conductor by 1000 to derive a unit of measurement in mils, that is,

$$0.102 \times 1000 = 102 \text{ mils}$$

Using the formula, circular mils (cmil) area $= d^2$, where d is equal to 102 mils, the circular mils area of the conductor is,

$$\text{cmils area} = (102 \text{ mils})^2$$
$$= 10404 \text{ cmil}$$

The value (10380 cmil) listed in Table 8 differs slightly from the calculated value due to the actual diameter of the conductor being rounded off to the nearest whole number; where the square root of 10380 is 101.88.

25. A load being supplied by a branch-circuit is located 438 ft. from its source. If the branch circuit conductors are required to be 453 feet one-way with a .8 ohm maximum resistance per conductor, determine the resistance of the branch circuit conductors at 980 feet.

Before performing any calculations it is safe to conclude that the final answer will be larger than .8 ohm based on the fact that as the length of the conductors increases the conductor's resistance will do so likewise. With this in mind, the given resistance and length of the conductors yields a ratio of .8-to-453 where the resistance of the wire is equivalent to a specific length or expressed as .8 ohm/453 ft. Applying the given values of the ratio a proportional ratio can be established where,

$$\frac{.8 \text{ ohm}}{453 \text{ ft.}} = \frac{x}{980 \text{ ft.}}$$

with the variable "x" representing the unknown resistance value at 980 feet. By using the three given values the proportional ratio can be converted to find "x" where,

$$x = \frac{.8 \text{ ohm} \times 980 \text{ ft.}}{453 \text{ ft.}} = \frac{784 \text{ ohms}}{453} = 1.73 \text{ ohms}$$

Just as stated earlier the final answer is larger than the initial .8 ohm. Therefore, at 980 feet the resistance of the conductors results to 1.73 ohms.

For similar situations such as these the following formulas can be used to find either an unknown resistance value or an unknown conductor length.

<u>To determine unknown resistance</u>   <u>To determine unknown length</u>

$$R_x = \frac{R}{L_1} \times L_2 \qquad\qquad L_x = \frac{L}{R_1} \times R_2$$

26. Determine the approximate length of a 6 AWG aluminum conductor that has a DC resistance of .035 ohm. Determine the approximate length of a 6 AWG uncoated copper conductor that has a DC resistance of 0.04 ohm.

Per Table 8, a 6 AWG aluminum conductor has a rated resistance of .808 ohm/kFT (.808 ohm per 1000 FT) or vice-versa, 1000 FT at .808 ohm. Applying the derived formula in question No. 25. the approximate length of the conductor is so determined:

$$L_x = \frac{1000\text{ft.}}{.808\ \cancel{\text{ohm}}} \times .035\ \cancel{\text{ohm}} = 43.32\text{ft.}$$

The approximate length of the 6 AWG aluminum conductor is 43.32ft.

Per Table 8, a 6 AWG uncoated copper conductor has a rated resistance of .491 ohm/kFT. Applying the derived formula the approximate length of the conductor is so determined:

$$L_x = \frac{1000\text{ft.}}{.491\ \cancel{\text{ohm}}} \times .04\ \cancel{\text{ohm}} = 81.47\text{ft.}$$

The approximate length of the 6 AWG uncoated copper conductor is 81.47ft.

27. Calculate the total DC resistance of three parallel 500 kcmil AL conductors. (The total length of each conductor is 1000 feet.)

Per Table 8, a 500 kcmil AL conductor has a rated resistance of .0424 ohm/kFT. Because the conductors are in parallel the equal resistance method for calculating parallel resistance can be applied to determine the resistance. As a result,

$$R = \frac{.0424\text{ ohm}}{3} = .0141\text{ ohm}$$

The total DC resistance of the three parallel conductors is .0141 ohm.

28. What is the resistance of an uncoated 8 AWG stranded copper conductor at 67°C and a 2 AWG stranded aluminum conductor at 83°C?

Table 8 lists the resistance of an uncoated 8 AWG stranded copper conductor at .778 ohm/1000 ft. and a 2 AWG stranded aluminum conductor at .319 ohm/1000 ft. based on a direct-current resistance at 75°C(167°F). To determine the resistance of both conductors at the requested temperatures, an equation is provided in **Note 2** of Table 8 for adjusting the resistance of a conductor when exposed to a temperature other than 75°C. Using the formula,

$$\mathbf{R_2 = R_1[1 + \alpha(T_2 - 75)]}$$

where,
$\mathbf{R_2}$ = Adjusted Resistance, $\mathbf{R_1}$ = Resistance at **75°C**, T_2 = Adjusted Temperature
$\mathbf{\alpha_{CU}}$ = 0.00323 and $\mathbf{\alpha_{AL}}$ = 0.00330 at **75°C**
α Greek letter Alpha = Temperature coefficient of resistance for the conductor material

Resistance (R_2) of uncoated 8 AWG stranded copper conductor at 67°C
R_2 = .778 ohm/1000ft. x [1 + 0.00323(67 − 75)]
= .778 ohm/1000ft. x [1 + 0.00323(-8)]
= .778 ohm/1000ft. x [1 + (-0.02584)]
= .778 ohm/1000ft. x [.97416]
= .758 ohm/1000ft. *(rounded off)*

Resistance (R_2) of 2 AWG aluminum conductor at 83°C
R_2 = .319 ohm/1000ft. x [1 + 0.00330(83 − 75)]
= .319 ohm/1000ft. x [1 + 0.00330(8)]
= .319 ohm/1000ft. x [1 + (0.02640)]
= .319 ohm/1000ft. x [1.02640]
= .327 ohm/1000ft. *(rounded off)*

Based on the calculated results for both conductors the relationship between the temperature and the resistance of the conductors is proportional. As the temperature decreases, the resistance decreases and vice-versa.

Table 9 - Alternating-Current Resistance and Reactance for 600-Volt Cables, 3-Phase, 60Hz, 75°C (167°F) - Three Single Conductors in Conduit (29. - 33.) [5]

DC *versus* AC resistance

When the total cross-sectional area of a conductor is consumed, the electrical resistance is known as the "DC resistance". In a direct current (dc) circuit, electrical current flows at uniform (even) density (the area in relationship to its size) throughout the entire cross-sectional area (diameter) of a conductor. See Figure CH9-9.

DC vs AC RESISTANCE

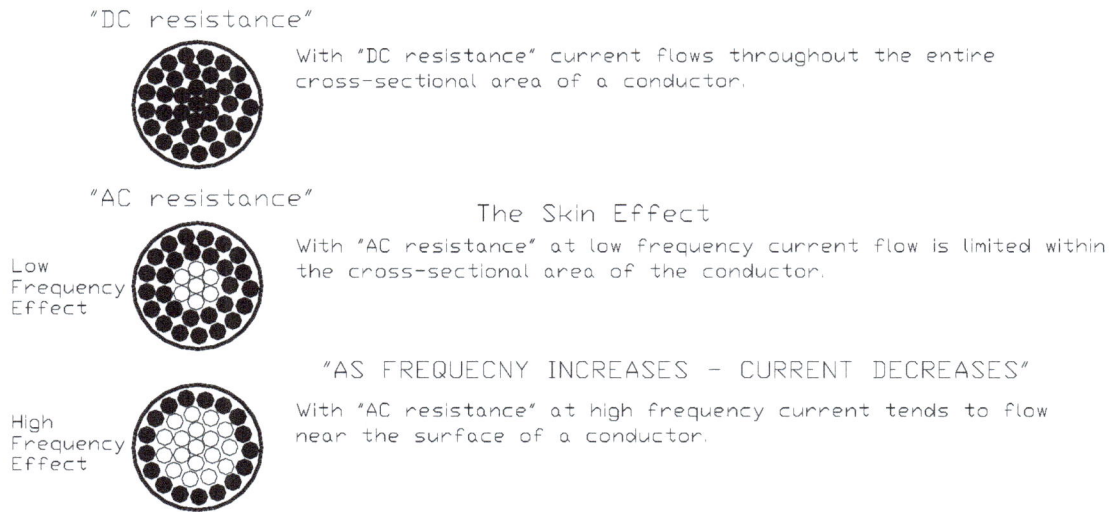

"DC resistance"

With "DC resistance" current flows throughout the entire cross-sectional area of a conductor.

"AC resistance"

Low Frequency Effect

The Skin Effect

With "AC resistance" at low frequency current flow is limited within the cross-sectional area of the conductor.

"AS FREQUECNY INCREASES – CURRENT DECREASES"

With "AC resistance" at high frequency current tends to flow near the surface of a conductor.

High Frequency Effect

Figure CH9-9

For electrical conductors, this uniform density is evaluated in terms of current flow per cross-sectional area which is known as the current density. In quantity, current density is expressed in units of amperes per meters squared or amperes per millimeters squared (A/m^2 or A/mm^2).

Unlike *dc* circuits, electrical current does not tend to flow in uniform density throughout the cross-sectional area of a conductor in an *ac* circuit, especially for larger size conductors. As the size of a conductor increases so does its cross-sectional area. In an *ac* circuit, electrical current does not flow uniformly throughout the cross-sectional area of a conductor but generally near the surface or outer layer of the conductor. This manner of current flow is known as the *skin effect*. The *skin effect* is where *ac* current tends to avoid flowing through the entire conductor but instead limits its conduction near the surface or outer layer. Skin effect becomes more evident as the frequency of an *ac* system is increased. The higher the frequency the more current is limited and restricted thus causing resistance in a conductor. This type of resistance is known as "AC resistance". Skin effect can be reduced by using stranded conductor's rather solid conductors. By using stranded conductors the effective surface area of a conductor is increased, particularly those conductors of larger sizes.

The *skin effect* is caused by *eddy currents* which are small independent currents that are produced by the effects of expanding and collapsing magnetic fields of *ac* current flow. Because eddy currents are much stronger about the center of a conductor they tend to force current to flow near a conductor's outer surface.

Conductors of sizes 1/0 AWG and larger are mostly effected by the *skin effect*. Therefore, when performing any type voltage drop calculations regardless of method most calculations will not provide as much accuracy as the methods prescribed in **Table 9** of **Chapter 9** and related notes. For voltage drop calculations containing direct-current (*dc*) resistance values refer to topic, **VOLTAGE DROP CALCULATIONS**.

Table 9 of **Chapter 9** provides a more effective method for performing basic voltage drop calculations involving *ac* circuits. The table list inductance reactance (X_L) and ac resistances (**R**) values for uncoated copper and aluminum wires ranging from 14 AWG to 1000 kcmil that are enclosed in three types of conduit (PVC, Aluminum, and Steel). Based on the bold printed heading of the table, these values are limited to 600-Volt cables operating at three-phase 60 Hz with a temperature rating of 75°C (167°F) where three single conductors are enclosed in one of the listed conduits. The table also list calculated effective impedance (Z) values at an 85 percent power factor. Where an *ac* circuit operates at a different power factor (PF) **Note 2** of Table 9 provides an equation to calculate the effective impedance (*Ze*). Applying a simplified version of the given equation, the effective impedance can also be calculated using the following equation:

$$Ze = (R \times \cos\theta) + (X_L \times \sin\theta) - where \cos\theta = PF \text{ and } \theta = PF \cos^{-1}$$

Per **Note 1**, capacitance reactance is ignored since it is deemed negligible (insignificant).

29. Determine the *ac* resistance of an uncoated 1/0 AWG copper conductor when 678 feet of the conductor is installed in aluminum conduit.

Per Table 9, an uncoated 1/0 AWG copper conductor in aluminum conduit has a rated resistance of .13 ohms to Neutral per 1000 feet. Applying the derived formula in question No. 25. the approximate *ac* resistance of the conductor is so determined:

$$R_x = \frac{.13 \text{ ohm}}{1000 \text{ ft.}} \times 678 \text{ ft.} = .08814 \text{ ohm}$$

The approximate resistance of the 1/0 AWG uncoated copper conductor installed in aluminum conduit is .08814 ohm.

30. Determine the *ac* resistance of a 250 kcmil aluminum conductor when 392 feet of the conductor is installed in steel conduit.

Per Table 9, a 250 kcmil aluminum conductor in steel conduit has a rated resistance of .086 ohms to Neutral per 1000 feet. Applying the formula the approximate *ac* resistance of the conductor is so determined:

$$R_x = \frac{.086 \text{ ohm}}{1000 \text{ ft.}} \times 392 \text{ ft.} = .033712 \text{ ohm}$$

The approximate resistance of the 250 kcmil aluminum conductor installed in steel conduit is .033712 ohm.

31. A 575V, 3-phase, 3-wire underground feeder is ran 1523 feet before terminating in a 600A motor control center. The feeder consists of 1000 kcmil THWN/THHN copper conductors with a temperature rating limited to 75°C. If the feeder was installed in 6" PVC-Schedule 40 and operate at an 85 percent (.85) power factor how much *voltage drop* would the feeder conductors encounter?

The data given in Table 9 is based on the operating conditions listed in the table's heading. The heading reference the use of alternating current (*ac*) resistance and reactance for 3-phase, 600 volt cables rated for 75°C (167°F) at a frequency of 60 hertz (Hz) for three single conductors in conduit.

Note 1 of Table 9 lists certain operating boundaries pertaining to the information and data given in Table 9.

Using the information found in the column titled "Effective Z at 0.85 PF for Uncoated Copper Wires" of Table 9 for PVC conduit, the effective impedance (Z) for a 1000 kcmil copper conductor is .032Ω. The impedance values listed in this table are based on Ohms to Neutral per 1000 feet (bottom row) or where metric units are required, Ohms to Neutral per Kilometer (top row).

This impedance value is based on 1000 feet and the length of the circuit is 1523 feet. The proportionalized ratio of the impedance per length along with the given length and the circuit load are multiplied together to derived the line-to-neutral voltage drop (R x I) therefore,

$$\frac{.032Ω}{1000ft.} \quad x \quad 1523ft. \quad x \quad 600A = 29.24V$$

To find the three-phase voltage drop the line-to-neutral voltage drop is multiplied by 1.732 ($\sqrt{3}$). As a result, the feeder conductors would encounter a voltage drop of 50.64V (29.24V x 1.732).

32. Utilizing the information given in question No. 31., use the given formulas below (refer to topic, **VOLTAGE DROP CALCULATIONS**) along with the applicable *dc* resistance values to calculate the voltage drop and compare.

Applying **THE RESISTIVE METHOD** the voltage drop is calculated using the following formula:

$$V_D = \frac{\sqrt{3} \; x \; R \; x \; L \; x \; I}{1000'}$$

According to Table 8, the *dc* resistance for a 1000 kcmil uncoated copper conductor is .0129 ohm. As a result,

$$V_D = \frac{\sqrt{3} \; x \; .0129Ω \; x \; 1523' \; x \; 600A}{1000'}$$
$$= 20.42V$$

Applying **THE CIRCULAR-MIL METHOD** the voltage drop is calculated using the following formula:

$$V_D = \frac{\sqrt{3} \; x \; K \; x \; L \; x \; I}{CM}$$

According to the given method, the *dc* resistance (K) for a copper conductor is 12.8 ohms; where 1000 kcmil is equal to 1,000,000 mils. As a result,

$$V_D = \frac{\sqrt{3} \times 12.8\Omega \times 1523' \times 600A}{1,000,000}$$
$$= 20.26V$$

Based on the calculated results the *dc* voltage drops proved to be lower than the *ac* voltage drop. The *ac* voltage drop is approximately 2.5 times (50.64V/20.42V - 50.64V/20.26V) that of the *dc* voltage drops. The results demonstrate the inaccuracy mentioned earlier where *ac* voltage drop is calculated applying *dc* parameters.

33. Re-calculate question No. 31. if the power factor was 78 percent (.78). If the power factor was 92 percent (.92).

According to **Note 2** of Table 9 the effective impedance values shown in Table 9 are valid only an 0.85 power factor. Where another circuit power factor (PF) other than 0.85 exists, the effective impedance (*Ze*) can be calculated applying the **R** and **X$_L$** values given in Table 9. Using the simplified formula:

$$Ze = (R \times \cos \theta) + (X_L \times \sin \theta)$$

Applying the 78 percent (.78) power factor

Before applying the effective impedance formula the circuit's phase angle (θ) has to be determined where,

$$\theta = .78 \cos^{-1}$$
$$= 38.74°$$

and per Table 9, 1000 kcmil copper in PVC (based on Ohms to Neutral per 1000 Feet) has a "R" value of .015Ω and a "X$_L$" value of .037Ω. Substituting the phase angle and resistance values in the effective impedance formula,

$$Ze = (.015 \times .78) + (.037 \times [\sin 38.74° = .626])$$
$$= .0117 \qquad .023162$$
$$= .034862 \text{ ohm to neutral}$$

The effective impedance of a 1000 kcmil copper conductor operating at a 78 percent power factor in PVC conduit is .034862Ω. At .034862Ω, the line-to-neutral voltage drop is,

$$\frac{.034862\Omega \times 1523ft. \times 600A}{1000ft.} = 31.86V$$

and the three-phase voltage drop is, 31.86V x 1.732 = 55.18V. At a power factor of 78 percent the feeder conductors would encounter a 55.18V voltage drop.

Applying the 92 percent (.92) power factor

Again, before applying the effective impedance formula the circuit's phase angle (θ) has to be determined where,

$$\theta = .92 \cos^{-1}$$
$$= 23.07°$$

Substituting the phase angle and resistance values in the effective impedance formula,

$$Ze = (.015 \times .92) + (.037 \times [\sin 23.07° = .392])$$
$$= .013 + .014504$$
$$= .028304 \text{ ohm to neutral}$$

The effective impedance of a 1000 kcmil copper conductor operating at a 92 percent power factor in PVC conduit is .028304Ω. At .028304Ω, the line-to-neutral voltage drop is,

$$\frac{.028304\Omega \times 1523\text{ft.} \times 600A = 25.86V}{1000\text{ft.}}$$

and the three-phase voltage drop is, 25.86V x 1.732 = 44.79V. At a power factor of 92 percent the feeder conductors would encounter a 44.79V voltage drop.

VOLTAGE DROP CALCULATIONS

As discussed earlier, line loss in an electrical conductor can cause dependent loads to respond inefficiently and shorten the life of connected equipment. Regardless of how many specialty calculators or quick-use references available in today's market, guaranteeing instant results, knowing how to apply and calculate voltage drop or voltage drop related applications is vital in the electrical profession. The NEC either recommends [210.19(A) Informational Note No. 4; 215.2(A) Informational Note No. 2; 310.15(A)(1) Informational Note No. 1, 455.6 Informational Note and 551.73(D) Informational Note] *or* require the application *or* the consideration of voltage drop in sections of the Code [647.4(D), 695.7, question Nos. 5. and 35. of Article 250 (Volume 2)].

The formulas given in each method below is an extension to that which has already been discussed and illustrated. Both methods will be used to perform *voltage drop* **(V$_D$)** calculations pertaining to *single- and three-phase circuits*. These formulas will also be used to derive other useful formulas to determine *the resistance* **(R),** *circular mils* **(CM)**, *maximum length* **(L)** or *maximum current (load)* **(I)** that's relevant to circuit conductors. Before proceeding review the formulas of each method for clarity and application. For the purpose of examination either method will provide needed results.

Because most voltage drop calculations are performed applying direct-current (*dc*) resistance values, either of the following methods can be applied. Although both methods are useful for calculating voltage drop based on *dc*-resistance, the results are not nearly as accurate when used for alternating-current (*ac*) circuits. Again, for more advanced techniques and practical examples pertaining to the use of *ac* resistance (and reactance) see question Nos. 29. through 33.

THE RESISTIVE METHOD

$$V_D = \frac{2 \times R \times L \times I}{1000} \quad \textit{(Single-phase circuits)}$$

$$V_D = \frac{2 \times R \times L \times I \times .866}{1000} \textit{ (Three-phase circuits) or } V_D = \frac{\sqrt{3} \times R \times L \times I}{1000}$$

where, V_D = Voltage drop in circuit voltage (V)
2 = Number of conductors considered in single-phase circuit
R = Direct-current (DC) resistance at 75°C for solid or stranded uncoated copper conductors or aluminum (ohms[Ω]/KFT)
<**Table 8** of **Chapter 9** - NEC>
L = One-way length of conductor (ft.)
I = Actual current (A) flow through conductor
1000 = Resistance of a given conductor per 1000 feet

THE CIRCULAR-MIL METHOD

$$V_D = \frac{2 \times K \times L \times I}{CM} \quad \textit{(Single-phase circuits)}$$

$$V_D = \frac{2 \times K \times L \times I \times .866}{CM} \quad \textit{(Three-phase circuits)} \text{ or } V_D = \frac{\sqrt{3} \times K \times L \times I}{CM}$$

where, V_D = Voltage drop in circuit voltage (V)
2 = Number of conductors considered in single-phase circuit
K = Direct-current (DC) resistance constant of conductor at 75°C
[Copper = 12.8Ω   Aluminum = 21.1Ω]
L = One-way length of conductor (ft.)
I = Actual current flow through conductor (A)
CM = Area of conductor in circular mils <**Table 8** of **Chapter 9** - NEC>

Note: 2 x .866 = 1.732 *or* the $\sqrt{3}$. The DC resistance constant (K) values used in this method for copper and aluminum conductors are not fixed and will vary depending upon reference material.

Although both formulas are useful for calculating voltage drop based on *dc*-resistance they are not accurate when pertaining to alternating-current *(ac)* circuits. For related formulas, questions, and examples pertaining to *voltage drop* based on *dc*-resistance the following provisions are provided.

SINGLE-PHASE (1ϕ) VOLTAGE DROP CALCULATIONS
(34. - 36.) [3]

34. Two 12 AWG THWN stranded copper conductors are used to supply a 1 horsepower motor that's rated for 115V. A 120 volt source located at a distance of 147 feet supplies the motor. Determine the *voltage drop* experienced by the conductors.

To determine the *voltage drop* for this installation the **Resistive** and **Circular-Mil** Methods will be used.

$$V_D = \frac{2 \times R \times L \times I}{1000'} \qquad V_D = \frac{2 \times K \times L \times I}{CM}$$

Using the information given in the question the following data is provided to determine the *voltage drop* applying the **Resistive Method**.

$$R = 1.98 \text{ ohms } (\Omega)/kFT \quad L = 147ft$$

The only data not given directly in the question is the amount of current (**I**). However, we do know that once the voltage source is placed across the motor, the motor will pull a certain amount of current that will flow through the 12 AWG conductors. Therefore, the only thing left to do now is to determine the amount. **Table 430.248** lists the full-load current for a 115 volt, 1 horsepower motor at 16A. With this final piece of data, the *voltage drop* can now be calculated.

Starting with the **Resistive Method**,

$$\begin{matrix} \mathbf{(R)} & & \mathbf{(I)} \end{matrix}$$
$$V_D = \frac{2 \times 1.98\Omega \times 147\text{ft} \times 16A}{1000\text{ft}}$$

$$= \frac{9313.92V}{1000} = 9.31V \text{ (rounded off)}$$

The calculated *voltage drop* is 9.31 volts.

Before moving on let's look at how the unit (of measurement) (**V**) following the value *9313.92* was determined. Notice that in the numerator (top part) of the formula there is a resistance value and a current value. Remember from **Ohm's Law** that **R x I = E**, where **E** represents the symbol for *volts*. In general, this formula is no more than an extension of Ohm's Law which takes the length of a conductor into consideration. Also notice that in the numerator and denominator there are units given in feet (ft); because they are like units they cancel each other out. This explains why the value 1000 is left without a unit in the denominator.

Now let's apply the **Circular-Mil Method**,

Since **L** and **I** have already been identified, the data needed for **K** and **CM** must be determined. Because stranded copper conductors are being used, **K** and **CM** are determined based on the referenced information listed below the formula.

Therefore,

$$\mathbf{K} = 12.8\Omega \qquad \mathbf{CM} = 6530$$

and when substituted into the formula the results are as follows:

$$V_D = \frac{2 \times 12.8\Omega \times 147\text{ft} \times 16A}{6530}$$

$$= \frac{60,211.2V}{6530} = 9.22V \text{ (rounded off)}$$

Unlike the detailed explanation given when applying the **Resistive Method**, the same approach for certain parts of the **Circular-Mil Method** is not as clear-cut. Comparing the results of both calculations, the differences are considered insignificant. When the motor is operating at its full load (16A) the operating voltage at the motor according to the results of the **Resistive Method** would be **110.69V** (120V − 9.31V) and according to the results based upon the **Circular-Mil Method, 110.78V** (120V − 9.22V), a difference of **.09V**. Keep in mind that each formula only provides a means for approximating results that are otherwise unavailable without such methods.

The final choice as to which method one may use is left up to the individual, some favor one over the other for different reasons.

35. Calculate the *voltage drop* experienced when a set of stranded 8 AWG THHN copper conductors supply a 120V-35A load. The load is located approximately 93 feet from an overcurrent device that's rated for 75°C.

Using the **Resistive Method**

$$\mathbf{R} = .778 \text{ ohms } (\Omega)/\text{kFT} \quad \mathbf{L} = 93\text{ft} \quad \mathbf{I} = 35\text{A}$$

$$\mathbf{V_D} = \frac{2 \times .778\Omega \times 93\text{ft} \times 35\text{A}}{1000\text{ft}} = \frac{5064.78\text{V}}{1000} = 5.06\text{V} \text{ (rounded off)}$$

The calculated voltage drop using the **Resistive Method** is 5.06 volts.

Using the **Circular-Mil Method**

$$\mathbf{K} = 12.8\Omega \quad \mathbf{L} = 93\text{ft} \quad \mathbf{I} = 35\text{A} \quad \mathbf{CM} = 16510$$

$$\mathbf{V_D} = \frac{2 \times 12.8\Omega \times 93\text{ft} \times 35\text{A}}{16510} = \frac{83,328\text{V}}{16510} = 5.05\text{V} \text{ (rounded off)}$$

The calculated voltage drop using the **Circular-Mil Method** is 5.05 volts.

36. A single-phase, 3W, 208V feeder is stubbed up in a maintenance shop that once fed an unknown load. The feeder consists of 3-2/0 AWG THW-2 aluminum conductors. The shop foreman desires to use the feeder conductors to temporarily supply a sub-panelboard that will be protected by a 175A overcurrent device located at least 210 feet away. Determine the *voltage drop* about the feeder conductors considering worst case.

At worst, the 2/0 AWG aluminum conductors could carry a load up to the rating of the 175A overcurrent device, therefore this value will be used for **I**.

Using the **Resistive Method**

$$\mathbf{R} = .159 \text{ ohms } (\Omega)/\text{kFT} \quad \mathbf{L} = 210\text{ft} \quad \mathbf{I} = 175\text{A}$$

$$\mathbf{V_D} = \frac{2 \times .159\Omega \times 210\text{ft} \times 175\text{A}}{1000\text{ft}}$$

$$= \frac{11,686.5\text{V}}{1000} = 11.7\text{V} \text{ (rounded off)}$$

At worst, the calculated *voltage drop* using the **Resistive Method** is 11.7 volts.

Using the **Circular-Mil Method**

$$K = 21.1\Omega \quad L = 210ft \quad I = 175A \quad CM = 133100$$

$$V_D = \frac{2 \times 21.1\Omega \times 210ft \times 175A}{133100}$$

$$= \frac{1,550,850V}{133100} = 11.65V \; \textit{(rounded off)}$$

At worst, the calculated *voltage drop* using the **Circular-Mil Method** is 11.65 volts.

Calculating Resistance (R) and Circular Mils (CM) (37. - 40.) [4]

When electrical conductors in an electrical system are installed in long runs, be it for service, feeder or branch circuits, *voltage drop losses* must be considered. When the amount of *voltage drop* in electrical conductors is excessive, conductors can overheat due to too much resistance or impedance causing insulation damage, poor power quality and other adverse operating conditions that could eventually lead to extreme hazards. To eliminate such conditions, *voltage drop* calculations should be performed. Although *voltage drop* calculations are not *NEC* enforceable, the *NEC* does address this concern through practical recommendations.

To maintain the *voltage drop* recommended by **NEC 210.19(A) Informational Note No. 4** for *branch circuits* and **NEC 215.2(A) Informational Note No. 2** for feeder circuits, the formulas of both methods can be converted to determine the appropriate size conductors needed to reduce or obtain a specific *voltage drop*. Since **resistance** and **circular mils** are the two main factors in each formula that's relative to the sizing of a conductor, both formulas will be converted to determine the needed *resistance* or *circular mils* of a conductor to *minimize voltage drop losses*.

Resistive Method

$$V_D = \frac{2 \times R \times L \times I}{1000'} \; \textit{(Original formula)} \qquad R = \frac{V_D \times 1000'}{2 \times L \times I} \; \textit{(Converted formula)}$$

Circular-Mil Method

$$V_D = \frac{2 \times K \times L \times I}{CM} \; \textit{(Original formula)} \qquad CM = \frac{2 \times K \times L \times I}{V_D} \; \textit{(Converted formula)}$$

37. To maintain a 3 percent voltage drop, what *size* conductors are required in questions 34. - 36.?

In question No. 34., 12 AWG conductors were used to supply the load at 120 volts, the source voltage. The desired 3 percent (.03) voltage drop is determined based on the source voltage; not the motor's rated voltage. Therefore, the desired voltage drop (V_D) is,

$$120V \times .03 = 3.6V$$

Substituting this value along with the other values into the converted formulas, the *resistance* and *circular mils* can be determined to *size* the needed conductors.

$$R = \frac{3.6V \times 1000'}{2 \times 147ft \times 16A} = .765\Omega \text{ (rounded off)}$$

with (E) above numerator and (I) below denominator.

Notice how the formula reveals $R = E / I$. Now let's determine **CM**.

$$CM = \frac{2 \times 12.8\Omega \times 147ft \times 16A}{3.6V} = 16725$$

Refer to **Table 8** of **Chapter 9**.

Resistance (R)

The calculated value for **R** will be used to select a conductor based on the *resistance* value gathered from the column which identifies *"uncoated copper"* conductors in ohm(s) per thousand feet (ohm/KFT) and the last column which identifies *"aluminum"* conductors in likewise units. The *resistance* value listed in either column when compared to the calculated value must be either *equal to* or *smaller than* the calculated value when selecting a conductor.

Since the calculated value for **R** is .765Ω, the *resistance* value listed in the column which reference *"uncoated copper conductors"* must be either .765Ω or smaller. Because the selected conductors will be *stranded* and .765Ω is not a standard resistance value, the *resistance* value listed in the column that is smaller than .765Ω is .491Ω, which reference a 6 AWG copper conductor. Although there is a slight difference between the resistance value for an 8 AWG stranded copper conductor, .778Ω and the calculated value, for examination purposes if using this formula, the 6 AWG conductor would be the correct choice. Just remember, the desired voltage drop **(V$_D$)** when based on a given percentage is dependent upon the source voltage. As the source voltage changes, so will the desired voltage drop. If the resistance value for the 8 AWG copper conductor (.778Ω) was substituted into the original formula, the results would be insignificant when compared to 3.6V, the desired voltage drop based on 120V. Observe,

$$V_D = \frac{2 \times .778\Omega \times 147ft \times 16A}{1000ft} = 3.66V \text{ (rounded off)}$$

Circular-Mils (CM)

When selecting a conductor based on the calculated value for **CM**, refer to the column of the table that reference *"Area-circular mils"*. The *circular mils* value listed in that column when compared to the calculated value must be either *equal to* or *greater than* the calculated value when selecting a conductor.

Since the calculated value for **CM** is 16725, the *circular mils* value listed in the column must be either 16725 or greater. Since 16725 is not a standard circular mils value, the value listed in the table that is greater than 16725 must be selected. Therefore, 26240 circular mils is selected which reference a 6 AWG copper conductor.

As you can see both methods yielded the same results. Since the process for sizing the conductors in question No. 34. involved detailed explanations based on the use of both converted formulas, the application of question Nos. 35. and 36. are limited to answers only.

Question No. 35.

Source voltage = 120V V_D = 120 x .03 = 3.6V

$$R = \frac{3.6V \times 1000'}{2 \times 93ft \times 35A} = .553\Omega \text{ (rounded off)}$$

(R) Per **Table 8** - The *resistance* value listed in the column that is *smaller* than .553Ω is .491Ω, which reference a 6 AWG copper conductor.

$$CM = \frac{2 \times 12.8\Omega \times 93ft \times 35A}{3.6V} = 23147 \text{ (rounded off)}$$

(CM) Per **Table 8** - The *CM* value listed in the column that is *greater* than 23147 is 26240, which reference a 6 AWG copper conductor.

Question No. 36.

Source voltage = 208V V_D = 208 x .03 = 6.24V

$$R = \frac{6.24V \times 1000'}{2 \times 210ft \times 175A} = .0849\Omega \text{ (rounded off)}$$

(R) Per **Table 8** - The *resistance* value listed in the column that is *smaller* than .0849Ω is .0847Ω, which reference a *250 kcmil* aluminum conductor.

$$CM = \frac{2 \times 21.2\Omega \times 210ft \times 175A}{6.24V} = 249712 \text{ (rounded off)}$$

(CM) Per **Table 8** - The *CM* value listed in the column that is *greater* than 249712 is 250000, which reference a *250 kcmil* aluminum conductor also.

38. A single-phase 37.5kVA transformer has a 240/120V secondary rating. The transformer will feed a 200A fusible disconnect switch (fused at 150A-75°C terminals) located 15' away. The disconnect switch serves as the disconnecting means for a main panelboard that's located

117' from the disconnect switch. Determine the minimum *size* THWN copper conductors needed to separately supply the disconnect switch and panelboard. (Limit all voltage drops to 2 percent.)

First let's consider the transformer's secondary current,

$$\mathbf{I} = \frac{37.5kVA}{240V} = 156.3A \text{ (rounded off)}$$

Although the transformer's secondary current is 156.3A, the actual load that will be placed on the conductors supplying the disconnect switch from the transformer is limited to the rating of the overcurrent device, 150A.

At a distance of 15 feet away from the transformer the conductors supplying the disconnect switch will be determined based on a voltage drop of 2 percent. Now the only thing left to do before proceeding is to determine the voltage drop $(\mathbf{V_D})$ at 2 percent as required.

$$V_D = 240V \text{ x } .02 = 4.8V$$

Applying both formulas,

$$\mathbf{R} = \frac{4.8V \text{ x } 1000'}{2 \text{ x } 15ft \text{ x } 150A} = 1.067\Omega \text{ (rounded off)}$$

$$\mathbf{CM} = \frac{2 \text{ x } 12.8\Omega \text{ x } 15ft \text{ x } 150A}{4.8V} = 12000$$

Per **Table 8** - The *resistance* value listed in the column that is *smaller* than 1.067Ω is $.764\Omega$. The *CM* value listed in the column that is *greater* than 12000 is 16510. Based on the results of both calculations, an 8 AWG copper conductor is referenced which according to **Table 310.15(B)(16)** has an ampacity of 40A at 75°C. Obvious there's a problem here. An 8 AWG conductor under no conditions could be used to supply a 150A load. The problem here lies within the length of the circuit. At 15 feet, the circuit's resistance is overwhelmingly reduced compared to the resistance values listed in **Table 8** which are all based on 1000 feet. At 1000 feet, 15 feet represents only 1.5 percent [15/1000] (.015) of the overall distance. Based on this small percentage value the resistance of a conductor would be limited likewise. Without a doubt, a distance of 15 feet will not cause a concern for voltage drop less alone a voltage drop calculation.

This problem serves to prove that although both formulas were used correctly the calculated results proved to be inaccurate and in some situations can't be used at all.

How do you derive an answer for this problem? Well, first *size* the conductor to the load and then calculate the voltage drop based on the resistance of the conductor. In other words, a 1/0 AWG copper conductor rated for 75°C has an ampacity of 150A which is identical to the load. Per **Table 8** the resistance of a 1/0 AWG "uncoated copper conductor" is $.122\Omega$ per 1000 feet.

Using this resistance value and the other given parameters, the 15 feet circuit will experience the following voltage drop,

$$V_D = \frac{2 \times .122\Omega \times 15' \times 150A}{1000'} = .549V$$

At .549V the circuit will only experience a .23 percent voltage drop (.549V/240V) which is well below the 2 percent voltage drop requirement (4.8V). At 1.5 percent the resistance of the conductor is,

$$.122\Omega \times .015 = .0018\Omega$$

At .0018Ω the conductor will encounter very little resistance which contributes to such a small drop in voltage.

Another means of approaching this situation is to calculate the required length of the conductors at the corresponding resistance (.122Ω) and ampacity of the conductors (150A) based upon the load current to determine a length that will yield the desired voltage drop. Observe,

$$L = \frac{4.8V \times 1000'}{2 \times .122\Omega \times 150A} = 131' \text{ (rounded off)}$$

$$L = \frac{4.8V \times 105600\ CM}{2 \times 12.8\Omega \times 150A} = 132'$$

Considering the calculated length, in order to derive a resistance or circular-mils that's consistent with the desired voltage drop (4.8V) the distance from the transformer's secondary to the disconnect switch would require being a minimum of 131 feet away, which is approximately the total length (15' + 117') of the entire installation. With these results everything could stop here, concluding that 1/0 AWG conductors could be used for both runs of the installation however, the question requires a further step. Needless to say, because the original resistance formulas did not provide a reasonable conclusion others could. In later parts of this topic (*Voltage Drop Calculations*), length and other derived components will be covered.

Finally, and only because it was a part of the question, the voltage drop between the disconnect switch and the panelboard must be determined based on the 117 feet distance and the 2 percent voltage drop. Applying both formulas,

$$R = \frac{4.8V \times 1000'}{2 \times 117ft \times 150A} = .137\Omega \text{ (rounded off)}$$

(R) Per **Table 8** - The *resistance* value listed in the column that is *smaller* than .137Ω is again, .122Ω, which reference a 1/0 AWG copper conductor.

$$CM = \frac{2 \times 12.8\Omega \times 117ft \times 150A}{4.8V} = 93600$$

(CM) Per **Table 8** - The *CM* value listed in the column that is *greater* than 93600 is 105600, which reference a 1/0 AWG copper conductor also.

Again, the results yield the same conductor size that would satisfy both runs of the entire installation.

39. Refer to Figure 39. A 100A main lugs only (MLO) sub-panelboard is supplied from a main distribution panelboard (MDP). The sub-panelboard is protected by a 100A circuit breaker and located approximately 177' from the MDP. The voltage at the MDP measures 245/122V. A 120V single-phasse non-continuous load that's fed from the sub-panelboard is located 85' away. If the load pulls 28A when energized, what size copper feeder and branch circuit conductors are needed to supply the sub-panelboard and the 120V load respectively? (Use **NEC** recommendations)

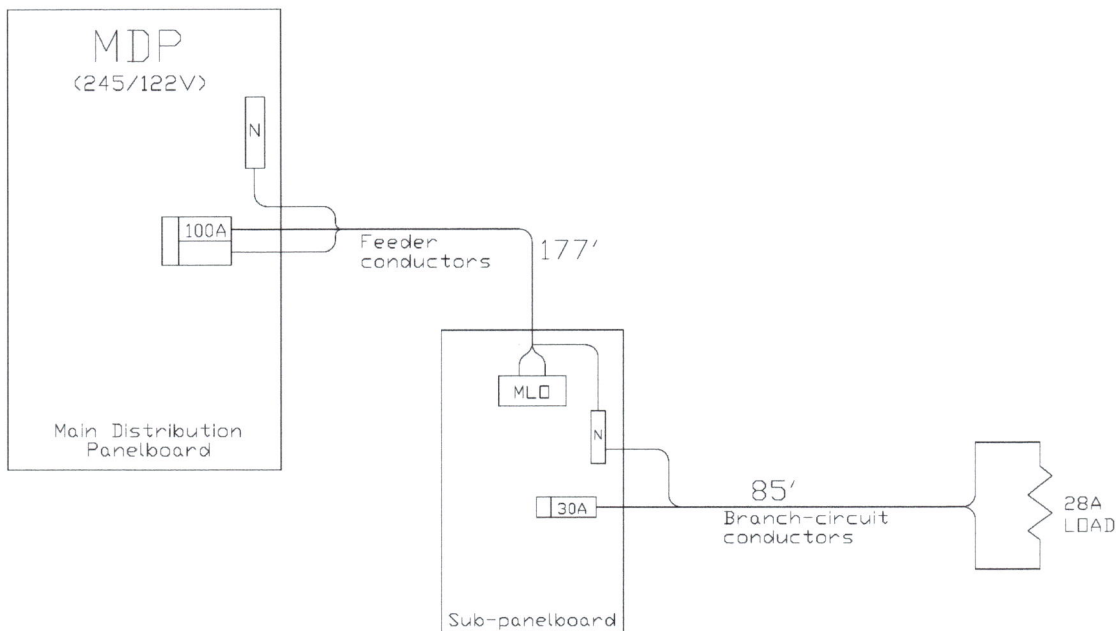

Figure 39

As recommended, NEC 210.19(A) **Informational Note No. 4** and NEC 215.2(A) **Informational Note No. 2** recommends the maximum *voltage drop* for *feeder* or *branch circuits* not exceed 3 percent and the maximum combined voltage drop for both feeder and branch circuits not exceed 5 percent. Since the voltage supply to the 120V-28A load is actually initiated from the MDP through the sub-panelboard, the total distance is 262'. At 5 percent of the 122V (122V x .05 = 6.1V) measured at the MDP, the voltage at the load when energized should have a voltage measurement no less than **115.9V** (122V − 6.1V).

Now let's begin by sizing the feeder conductors based on a 3 percent *voltage drop* of the feeder voltage (245V), followed by the branch circuit conductors, to determine if the voltage at the load is equal to or greater than 115.9V in order to maintain the recommended 5 percent voltage drop.

Feeder Conductors (From MDP to sub-panelboard – 100A maximum load)

$$V_D = 245V \times .03 = 7.35V$$

From the 100A feeder overcurrent device of the MDP, to the main lugs of the sub-panelboard, the voltage based on a 3 percent voltage drop would yield 237.65V (245V – 7.35V).

Applying both formulas to determine the size feeder conductors needed,

$$R = \frac{7.35V \times 1000ft}{2 \times 177ft \times 100A} = .2076\Omega$$

$$CM = \frac{2 \times 12.8\Omega \times 177ft \times 100A}{7.35V} = 61649 \text{ (rounded off)}$$

Based on the results of both calculations, the feeder conductors should be 2 AWG copper per **Table 8** of **Chapter 9**.

Branch Circuit Conductors (From sub-panelboard to neutral load – 28A load [120V])

$$\textbf{Neutral Voltage} = 237.65V/2 = 118.83V \text{ (rounded off)}$$
$$V_D = 118.83V \times .03 = 3.56V \text{ (rounded off)}$$

The voltage drop between the sub-panelboard and the load must be limited to 3.56V to maintain the maximum 5 percent voltage drop as recommended. At a voltage drop of 3.56V, the voltage at the load will result to 115.27V (118.83V - 3.56V) which is less than the initial calculated voltage, **115.9V**.

Applying both formulas,

$$R = \frac{3.56V \times 1000'}{2 \times 85ft \times 28A} = .7479\Omega$$

$$CM = \frac{2 \times 12.8\Omega \times 85ft \times 28A}{3.56V} = 17115 \text{ (rounded off)}$$

Based on the results of both calculations, the branch circuit conductors should be 6 AWG copper per **Table 8** of **Chapter 9**.

Therefore, in order to maintain a 5 percent voltage drop between the MDP and the 28A load, the feeder conductors must be **2 AWG** copper and the branch circuit conductors must be **6 AWG** copper.

40. Two unused 1 AWG THWN-2 aluminum conductors enclosed in a junction box are fed from a 208V-1ϕ source that's located 61' in distance. An electrician intends to join these

conductors with other aluminum conductors to supply a 208V, single-phase, 75A load installed 107' away. Determine the minimum *size* of the other conductors needed to supply the load.

The first thing to do is to determine the voltage drop of the aluminum conductors at 61' when the 75A load is applied. Refer to **Table 8** of **Chapter 9** for resistance and circular-mils values.

$$\mathbf{V_D} = \frac{2 \times .253\Omega \times 61' \times 75A}{1000'} = 2.31V \text{ (rounded off)}$$

$$\mathbf{V_D} = \frac{2 \times 21.1\Omega \times 61' \times 75A}{83690} = 2.31V \text{ (rounded off)}$$

Since the recommended *voltage drop* at 3 percent of the source voltage is 6.24V (208V x .03), the calculated results must be subtracted from the recommended voltage drop to determine the allowed voltage drop of the other conductors.

$$\text{Applying } \mathbf{R} \text{ (}V_D\text{) - } 6.24V - 2.32V \text{ (maximum)} = 3.92V$$

The allowed voltage drop can now be used to determine the size of the other conductors.

Applying both formulas,

$$\mathbf{R} = \frac{3.92V \times 1000ft}{2 \times 107ft \times 75A} = .2442\Omega$$

$$\mathbf{CM} = \frac{2 \times 21.1\Omega \times 107ft \times 75A}{3.92V} = 86392 \text{ (rounded off)}$$

Per **Table 8** of **Chapter 9**, if the size of the other conductors was based on **.2442Ω**, 1/0 AWG aluminum conductors would be required and likewise if based on **86392 CM.**

Calculating Length (L)

The *length* of a conductor can contribute to either an increase or decrease in voltage drop losses about a conductor. Using the formulas below will allow the maximum length of a conductor to be determined in order to maintain the NEC recommended voltage drop limits.

Resistive Method

$$V_D = \frac{2 \times R \times L \times I}{1000'} \text{ (Original formula)} \qquad L = \frac{V_D \times 1000'}{2 \times R \times I} \text{ (Converted formula)}$$

Circular-Mil Method

$$V_D = \frac{2 \times K \times L \times I}{CM} \text{ (Original formula)} \qquad L = \frac{V_D \times CM}{2 \times K \times I} \text{ (Converted formula)}$$

41. Two 4 AWG THW copper conductors are used to supply a 67A single-phase outdoor lighting load. If the conductors are being fed from a 240V source, *how far* can they be installed from the source and still maintain a 3 percent voltage drop?

Recommended voltage drop (V_D) of 240V source at 3 percent.

$$V_D = 240V \times .03 = 7.2V$$

Applying both formulas,

$$L = \frac{7.2V \times 1000'}{2 \times .308\Omega \times 67A} = 174' \; (rounded \; off)$$

$$L = \frac{7.2V \times 41740 \; CM}{2 \times 12.8\Omega \times 67A} = 175' \; (rounded \; off)$$

The length of the conductors cannot exceed 175 feet.

Calculating Current (I)

The amount of *current* that's allowed to flow through a conductor is another factor that can contribute to voltage drop losses. Quite often it is necessary to limit the load placed on conductors to obtain a desired voltage drop.

Resistive Method

$$V_D = \frac{2 \times R \times L \times I}{1000'} \; (Original \; formula) \qquad I = \frac{V_D \times 1000'}{2 \times R \times L} \; (Converted \; formula)$$

Circular-Mil Method

$$V_D = \frac{2 \times K \times L \times I}{CM} \; (Original \; formula) \qquad I = \frac{V_D \times CM}{2 \times K \times L} \; (Converted \; formula)$$

42. Two 2 AWG THW copper conductors were previously installed in buried raceway to feed an outdoor lighting *load* located 273' away from a 277V source. Initially 12 single-phase, 6A fixtures were to be fed from the conductors, however it has been determined that 17 fixtures are actually required to produce the desired lighting output. Can the existing conductors be used to supply the 17 fixtures and still maintain a 3 percent voltage drop?

Recommended voltage drop (V_D) of 277V source at 3 percent.

$$V_D = 277V \times .03 = 8.31V$$

Applying both formulas,

$$\mathbf{I} = \frac{8.31V \times 1000ft}{2 \times .194\Omega \times 273ft} = 78.5A \ (\textit{rounded off})$$

$$\mathbf{I} = \frac{8.31V \times 66360}{2 \times 12.8 \ \Omega \times 273ft} = 79A \ (\textit{rounded off})$$

To maintain a 3 percent voltage drop, the load placed on the conductors must be limited to no more than 79A. To supply 17 fixtures, the load on the conductors would experience 102A (17 x 6A). Although the ampacity of a 2 AWG THW copper conductor per Table 310.15(B)16 is 115A, the conductors still cannot be used not only because of the voltage drop/load limitations but also because the 102A load must be considered continuous since it will supply outdoor lighting, which is expected to operate for 3 hours or more. At 127.5A (102A x 1.25), as a minimum 2/0 AWG THW conductors are required (calculate for **R** - use 8.31V, 127.5A, 273 ft.).

THREE-PHASE (1φ) VOLTAGE DROP CALCULATIONS

When calculating the voltage drop (V_D) of a three phase load the only difference between that of a single phase load is the square root of three $\sqrt{3}$ *or* 1.732, if rounded off to the nearest thousandths for more accuracy. Similar to the various formulas and problems used to calculate voltage drop and other relative factors for single phase loads, the same approach will also be used for three phase loads. Because the section on single-phase voltage drop provides an array of general details and information, thus relevant for use with three-phase voltage drop calculations also, such provisions are limited in this section. Again, before proceeding review the formulas of each method for clarity and application.

Three-phase circuits

Resistive Method

$$V_D = \overset{(1)}{\frac{\sqrt{3} \times R \times L \times I}{1000'}} \ or \ \overset{(2)}{\frac{1.732 \times R \times L \times I}{1000'}} \ or \ \overset{(3)}{\frac{2 \times .866 \times R \times L \times I}{1000'}}$$

Note: Because $\sqrt{3}$ = 1.732 = 2 x .866, any formula of either method can be used to obtain the same results.

where, V_D = Voltage drop in circuit voltage (V)
R = Direct-current (DC) resistance at 75°C for solid or stranded uncoated copper or aluminum conductors (ohms[Ω]/KFT)
    <Table 8 of **Chapter 9** - NEC>
L = One-way length of conductor (ft.)
I = Actual current (A) flow through conductor
1000' = Resistance of a given conductor per 1000 feet

Circular-Mil Method

$$V_D = \underbrace{\frac{\sqrt{3} \times K \times L \times I}{CM}}_{(1)} \quad or \quad \underbrace{\frac{1.732 \times K \times L \times I}{CM}}_{(2)} \quad or \quad \underbrace{\frac{2 \times .866 \times K \times L \times I}{CM}}_{(3)}$$

where, V_D = Voltage drop in circuit voltage (V)
K = Direct-current (DC) resistance constant of conductor at *75°C*
(Copper = 12.8Ω   Aluminum = 21.1Ω)
L = One-way length of conductor (ft.)
I = Actual current flow through conductor (A)
CM = Area of conductor in circular mils <**Table 8** of **Chapter 9** - NEC>

43. How much *voltage drop* would the feeder conductors encounter at full load capacity if 1000 kcmil THHN/THWN aluminum conductors were used instead in question No. 32.? Use the procedure applied in question No. 31. for calculating *voltage drop* and compare the calculated results.

Again, using the information given in question No. 32. and the resistance value of the aluminum conductors per **Table 8** [.0212 ohms (Ω)] the voltage drop is calculated applying both methods.

Starting with the **Resistive Method**,

$$V_D = \frac{1.732 \times .0212Ω \times 1523ft \times 600A}{1000ft}$$

$$= \frac{33553.27V}{1000} = 33.55V \text{ (rounded off)}$$

The calculated *voltage drop* using **Resistive Method** is 33.55 volts.

Now let's apply the **Circular-Mil Method**,

Since **L** and **I** have already been identified, the data needed for **K** and **CM** must be determined per given references listed below the formula. Therefore,

K = 21.1Ω   **CM** = 1,000,000

and when substituted into the formula the results are as follows.

$$V_D = \frac{1.732 \times 21.1Ω \times 1523ft \times 600A}{1,000,000}$$

$$= \frac{33395003.76V}{1,000,000} = 33.4V \text{ (rounded off)}$$

The calculated *voltage drop* using the **Circular-Mil Method** is 33.4 volts.

Per question No. 31.

Using the information found in the column titled, "Effective Z at 0.85 PF for Aluminum Wires" of Table 9 for PVC conduit, the effective impedance (Z) for a 1000 kcmil aluminum conductor is .039Ω.

This impedance value is based on 1000 feet and the length of the circuit is 1523 feet. The proportionalized ratio of the impedance per length along with the given length and the circuit load are multiplied together to derived the line-to-neutral voltage drop (R x I) therefore,

$$\frac{.039Ω}{1000\text{ft.}} \quad x \quad 1523\text{ft.} \quad x \quad 600A = 35.64V$$

To find the three-phase voltage drop the line-to-neutral voltage drop is multiplied by 1.732 ($\sqrt{3}$). As a result, the feeder conductors would encounter a voltage drop of 61.73V (35.64V x 1.732).

Based on the calculated results the *dc* voltage drops proved to be lower than the *ac* voltage drop. The *ac* voltage drop is approximately 1.85 times (61.73V/33.55V - 61.73V/33.4V) that of the *dc* voltage drops.

Calculating Resistance (R) and Circular Mils (CM) (44. - 45.) [2]

Resistive Method

$V_D = \frac{1.732 \times R \times L \times I}{1000'}$ *(Original formula)* $R = \frac{V_D \times 1000'}{1.732 \times L \times I}$ *(Converted formula)*

Circular-Mil Method

$V_D = \frac{1.732 \times K \times L \times I}{CM}$ *(Original formula)* $CM = \frac{1.732 \times K \times L \times I}{V_D}$ *(Converted formula)*

44. A 275kW load operating at a 90 percent power factor is located 138' from its 480V, 3-phase source. If XHHW copper conductors are used, what *size* conductors are required to maintain a 3 percent voltage drop?

The desired voltage drop (V_D) is,

$$480V \times .03 = 14.4V$$

To determine the load's operating current use the following formula,

$$\mathbf{I} = \frac{kW \times 1000}{E \times PF \times 1.732} \quad \text{where,} \quad \mathbf{I} = \frac{275 \times 1000}{480V \times .90 \times 1.732} = 367.54A$$

Substituting the known values into the converted formulas, the *resistance* and *circular mils* can be determined to size the needed conductors.

$$R = \frac{14.4V \times 1000ft}{1.732 \times 138ft \times 367.54A} = .164\Omega \text{ (rounded off)}$$

$$CM = \frac{1.732 \times 12.8\Omega \times 138ft \times 367.54A}{14.4V} = 78087 \text{ (rounded off)}$$

(Referring to) Table 8 of Chapter 9 indicates the need for the use of 1 AWG conductors which has a resistance (R) of .154Ω (uncoated copper) and a circular mils (CM) area of 83690.

Again, this situation serves to prove that although both formulas were used correctly the calculated results proved to be inaccurate and in some situations can't be used at all.

As implemented in a similar situation, the approach to deriving results that are satisfactory first requires the needed conductor to be sized to the load and then calculating the voltage drop based on the resistance of the conductor. As a minimum, according to Table 310.15(B)(16), a 500 kcmil XHHW conductor which has a rated ampacity of 380A will suffice. Per **Table 8** the resistance of a 500 kcmil "uncoated copper conductor" is .0258Ω per 1000 feet. Using this resistance value and the other given parameters, the 138 feet circuit will experience the following voltage drops per given methods,

$$V_D = \frac{1.732 \times .0258\Omega \times 138' \times 367.54A}{1000'} = 2.267V \text{ (rounded off)}$$

$$V_D = \frac{1.732 \times 12.8\Omega \times 138' \times 367.54A}{500000} = 2.249V \text{ (rounded off)}$$

At 2.267V/2.249V the circuit will only experience a .4723/.4685 percent voltage drop (2.267V/480V - 2.249V/480V) which is well below the required 3 percent voltage drop (14.4V).

45. What *size* aluminum conductors are needed in question No. 31. to reduce the voltage drop to 3 percent?

Recommended voltage drop (V_D) of 575V source at 3 percent.

$$V_D = 575V \times .03 = 17.25V$$

Applying both formulas,

$$R = \frac{17.25V \times 1000ft}{1.732 \times 1523ft \times 600A} = .0109\Omega \text{ (rounded off)}$$

$$CM = \frac{1.732 \times 21.1\Omega \times 1523ft \times 600A}{17.25V} = 1935942 \text{ (rounded off)}$$

Although the calculated results are slightly higher (.0109Ω) and a bit lower (1935942 CM or 1935.942 kcmil) compared to values listed for a 2000 kcmil *aluminum conductor* (.0106Ω – 2000000 CM or 2000 kcmil) in **Table 8** of **Chapter 9**, either this single conductor or two or more parallel conductors can be used which share an equivalent resistance of .0106Ω or less. Per Table 310.15(B)(16), the 2000 kcmil aluminum THWN/THHN conductors which are rated for 560A at 75°C are adequate to supply the 600A motor control center.

If parallel conductors were used in this case the actual combinations of required parallel conductors could be derived by using the *equal resistance value method*, where,

$$\mathbf{R_T} = \frac{\text{Individual resistance value}}{\text{Number of individual resistance values}}$$

In this case if two (2) **1000 kcmil aluminum conductors** were used instead, the total resistance between the two conductors would be,

$$\mathbf{R_T} = \frac{.0212\Omega}{2}$$

$$= .0106\Omega$$

Coincidentally, the results are the same as that for one 2000 kcmil aluminum conductor. Now consider the use of three (3) 700 kcmil aluminum conductors. The total resistance between the three conductors would be,

$$\mathbf{R_T} = \frac{.0303\Omega}{3}$$

$$= .0101\Omega$$

which is less than the previously calculated value .0109Ω. Nevertheless, if single 2000 kcmil aluminum conductors were used, a voltage drop resulting to 3 percent (17.25V) or less would be realized; observe,

$$\mathbf{V_D} = \frac{1.732 \times .0106\Omega \times 1523' \times 600A}{1000'} = 16.78V \text{ (rounded off)}$$

$$\mathbf{V_D} = \frac{1.732 \times 21.1\Omega \times 1523' \times 600A}{2000000} = 16.7V \text{ (rounded off)}$$

Calculating Length (L)

Resistive Method

$\mathbf{V_D} = \dfrac{1.732 \times R \times L \times I}{1000'}$ *(Original formula)* $L = \dfrac{V_D \times 1000'}{1.732 \times R \times I}$ *(Converted formula)*

Circular-Mil Method

$\mathbf{V_D} = \dfrac{1.732 \times K \times L \times I}{CM}$ *(Original formula)* $\mathbf{L} = \dfrac{V_D \times CM}{1.732 \times K \times I}$ *(Converted formula)*

46. At what *distance* can a 480V, 3-phase, 150kVA transformer be placed from its 460V source and maintain a 3 percent voltage drop, if supplied by 3/0 AWG copper conductors?

Recommended voltage drop (V_D) of 460V source at 3 percent.

$$V_D = 460V \times .03 = 13.8V$$

Other factors,

$$\mathbf{R} \text{ (3/0 AWG copper-uncoated)} = .0766 \text{ ohms}(\Omega)/kFT$$

$$\mathbf{CM} \text{ (3/0 AWG copper-uncoated)} = 167800$$

To determine **I**, calculate the transformer's full-load current using the following formula,

$$\mathbf{I} = \frac{kVA \times 1000}{E \times 1.732}$$

where,

$$\mathbf{I} = \frac{150 \times 1000}{460V \times 1.732} = 188.3A \text{ (rounded off)}$$

Applying both formulas,

$$\mathbf{L} = \frac{13.8V \times 1000'}{1.732 \times .0766\Omega \times 188.3A} = 552' \text{ (rounded off)}$$

$$\mathbf{L} = \frac{13.8V \times 167800 \text{ CM}}{1.732 \times 12.8\Omega \times 188.3A} = 555' \text{ (rounded off)}$$

The transformer can be distanced a maximum of 555 feet from its source which is totally dependent upon the desired voltage drop of the supplying conductors.

Calculating Current (I) (47. - 49.) [3]

Resistive Method

$$V_D = \frac{1.732 \times R \times L \times I}{1000'} \text{ (Original formula)} \qquad I = \frac{V_D \times 1000'}{1.732 \times R \times L} \text{ (Converted formula)}$$

Circular-Mil Method

$$V_D = \frac{1.732 \times K \times L \times I}{CM} \text{ (Original formula)} \qquad I = \frac{V_D \times CM}{1.732 \times K \times L} \text{ (Converted formula)}$$

47. An unidentified load is located 307' from its 208/120V, 3-phase voltage source. If 250 kcmil XHHW copper conductors are supplying the load, determine the allowable *load current* required to limit the voltage drop across the conductors to 3 percent.

Recommended voltage drop (V_D) of 208V source at 3 percent.

$$V_D = 208V \times .03 = 6.24V$$

Other factors,

$$\mathbf{R} \text{ (250 kcmil copper-uncoated)} = .0515 \text{ ohms}(\Omega)/kFT$$

$$\mathbf{CM} \text{ (250 kcmil copper-uncoated)} = 250000$$

Applying both formulas,

$$I = \frac{6.24V \times 1000ft}{1.732 \times .0515\Omega \times 307ft} = 228A \text{ *(rounded off)*}$$

$$I = \frac{6.24V \times 250000}{1.732 \times 12.8\Omega \times 307ft} = 229A \text{ *(rounded off)*}$$

The load can pull a maximum of 229A without compromising the desired 3 percent voltage drop.

48. If 250 kcmil XHHW aluminum conductors were used instead, determine the allowable *load current* required to limit the voltage drop across the conductors to 3 percent.

$$\mathbf{R} \text{ (250 kcmil aluminum)} = .0847 \text{ ohms } (\Omega)/kFT$$

$$\mathbf{CM} \text{ (250 kcmil aluminum)} = 250000$$

Applying both formulas,

$$I = \frac{6.24V \times 1000ft}{1.732 \times .0847\Omega \times 307ft} = 139A \text{ *(rounded off)*}$$

$$I = \frac{6.24V \times 250000}{1.732 \times 21.1\Omega \times 307ft} = 139A \text{ *(rounded off)*}$$

With the use of aluminum conductors, the load can pull a maximum of 139A without compromising the desired 3 percent voltage drop.

49. Refer to Figure 49. A 100A circuit supplied from a main distribution panel (MDP) rated for 208/120V-3ϕ feeds the 208/120V-1ϕ sub-panelboard located 256' away. The sub-panelboard is used to supply lighting for a designated area of a shopping mall's parking lot. The parking lot is illuminated with 15-250W paired metal halide fixtures where each fixture is rated for 1.73A at

208V-1φ. Three individual branch circuits will be used to feed each row of fixtures. What size feeder and branch circuit conductors (stranded copper-75°C) are required to supply the fixtures?

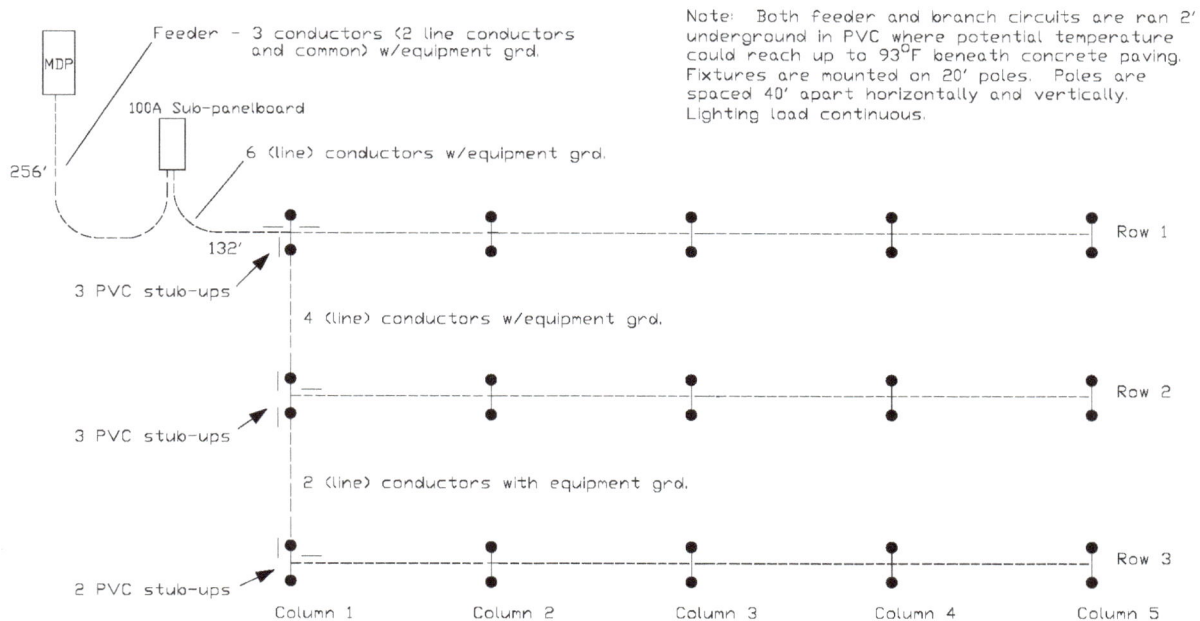

Figure 49

At the MDP (Supply voltage - 208V-1φ)

Considering both feeder and branch circuits, the total voltage drop will be limited to 5 percent, which is 10.4 volts (208V x .05). This limits the minimum supply voltage to 197.6 volts (208V – 10.4V). To realize the calculated minimum supply voltage, the feeder and branch-circuit voltage drop will be calculated at 3 and 2 percent, respectively. Once again, remember that the ambient temperature has the potential to reach up to 93°F and a 5' stub-up distance must be considered between each pole and used for the stub-up distance at the sub-panelboard.

Feeder (From MDP to 100A Panel)

Because the rating of the overcurrent device supplying the sub-panelboard is not given, the load current will be based upon the rating of the 100A sub-panelboard. Applying this and other information, resistance and circular-mils values can be calculated to determine the size feeder conductors needed.

$$V_D = 208V \times .03 = 6.24V$$
$$L = 261' \ (256' + 5')$$

$$R = \frac{6.24V \times 1000ft}{2 \times 261ft \times 100A} = .1195\Omega$$

$$CM = \frac{2 \times 12.8\Omega \times 261ft \times 100A}{6.24V} = 107077 \ (rounded \ off)$$

The calculated results for both formulas reference the use of 2/0 AWG copper conductors for the feeder. At 75°C, the conductors are rated for 175A and capable of supplying a continuous load.

Because the ambient temperature has the potential to reach up to 93°F the ampacity of the conductor has to be derated 94 percent (.94) per **Table 310.15(B)(2)(a)**. Therefore,

$$175A \times .94 = 164.5A$$

Even after being derated, this conductor still has the ampacity to carry a potential 100A load safely. Using a 2/0 AWG copper conductor and the resistance (.0967Ω) and circular-mils (133100 CM) values provided in **Table 8** of **Chapter 9**, the *voltage drop* from the MDP to the 100A sub-panelboard results to,

$$\mathbf{V_D} = \frac{2 \times .0967\Omega \times 261' \times 100A}{1000'} = 5.05V \text{ (rounded-up)}$$

$$\mathbf{V_D} = \frac{2 \times 12.8\Omega \times 261' \times 100A}{133100} = 5.02V \text{ (rounded off)}$$

Voltage at Sub-panelboard (applying worst case),

$$208V - 5.05V = 203V \text{ (rounded-up)}$$

Branch Circuits (BC)
The three branch circuits will be calculated individually from the 100A sub-panelboard. All calculated supply voltages will be referenced at each stub-up opposed to each fixture pair.

BC1 (Row 1)

To Column 1 - From sub-panelboard to the first stub-up (Row 1, Column 1), the referenced PVC per Figure 49 contains 6 current-carrying conductors. The branch-circuit conductors routed to this stub-up must be sized to carry the load [17.3A] of 10 fixtures a distance of 142' [5' + 132' + 5'] at a voltage drop not exceeding 2 percent of the voltage at the sub-panelboard (203V x .02 = 4.06V) to realize the calculated minimum supply voltage (197.6V) at the farthest fixture pair of Row 1. Therefore, to determine the size branch-circuit conductors needed, the following methods are applied:

$$\mathbf{R} = \frac{4.06V \times 1000'}{2 \times 142' \times 17.3A} = .8263\Omega \text{ (Use 8 AWG - [resistance .778}\Omega \text{ smaller than .8263}\Omega\text{])}$$

$$\mathbf{CM} = \frac{2 \times 12.8\Omega \times 142ft \times 17.3A}{4.06V} = 15490 \text{ (rounded off)} \text{ (reference use of 8 AWG)}$$

Based upon the selected use of 8 AWG conductors, because the conductors (two line-to-line) supplying the 10 fixtures will encounter an adverse ambient temperature while enclosed with four other current-carrying conductors (totaling six) both conditions must be factored in to determine whether these conductors are sufficient to carry the load. Per Tables 310.15(B)(2)(a) and 310.15(B)(3)(a), .94 and .80 are the factors that must be applied to determine the sufficiency of the conductors along with the provision for supplying a continuous load at 125 percent (1.25).

$$I = \frac{17.3A \times 1.25}{(.80 \times .94)} = 28.76A \text{ (ampacity less than 8 AWG @ 75°C [50A], 8 AWG o.k.)}$$

Now that the sufficiency of the conductors has been verified, the voltage drop that the conductors will experience from the sub-panelboard to the first stub-up can be calculated,

$$V_D = \frac{2 \times .778\Omega \times 142' \times 17.3A}{1000'} = 3.82V \text{ (rounded off - worst case to be applied)}$$

$$V_D = \frac{2 \times 12.8\Omega \times 142' \times 17.3A}{16510} = 3.81V \text{ (rounded up)}$$

which is used to determine the supply voltage of the branch-circuit conductors (BC1) at the first stub-up of Row 1.

$$203V - 3.82V = 199.18V$$

Now that the supply voltage at the *1st stub-up* has been determined, let's determine the supply voltage at the *last stub-up* (Column 5) using the same size 8 AWG branch-circuit conductors to determine whether the 5 percent limit between the feeder and branch circuit for this group of fixtures has been maintained. From here on out only one method will be used to calculate the various voltage drops where such voltage drops are rounded off where needed.

(Stub-up at Column 2 - from Column 1)

$$V_D = \frac{2 \times .778\Omega \times *50' \times 13.84A}{1000'} = 1.08V$$

$$199.18V - 1.08V = 198.1V \text{ (stub-up voltage)}$$

(Stub-up at Column 3 - from Column 2)

$$V_D = \frac{2 \times .778\Omega \times *50' \times 10.38A}{1000'} = .81V$$

$$198.1V - .81V = 197.29V \text{ (stub-up voltage)}$$

(Stub-up at Column 4 - from Column 3)

$$V_D = \frac{2 \times .778\Omega \times *50' \times 6.92A}{1000'} = .54V$$

$$197.29V - .54V = 196.75V \text{ (stub-up voltage)}$$

(Stub-up at Column 5 - from Column 4)

$$V_D = \frac{2 \times .778\Omega \times *50' \times 3.46A}{1000'} = .27V$$

$$196.75V - .27V = 196.48V \text{ (stub-up voltage)}$$

*length of stub-ups and distance between poles

Although the voltage drop at 196.48V falls below the *minimum supply voltage* (197.6V) it still is in reach of the 5 percent limitation. The difference between the initial supply voltage at 208V and 196.48V is 11.52V which results to an overall voltage drop of 5.54 percent (11.52V/208V). After the stub-up at column 2, the voltage drop fell below the minimum supply voltage and gradually declined. However, to meet or exceed the desired 5 percent limitation (197.6V) about the entire branch circuit (BC1), the 8 AWG conductors would require being replaced with 6 AWG conductors.

To finalize this series of calculations for this first branch circuit it is safe to conclude that the branch-circuit conductors from each stub-up to paired fixtures can be sized according to the load of the fixtures (3.46A). Depending upon the desired voltage loss, the conductors extending from each stub-up to the paired fixtures can be sized smaller than 6 AWG.

Now let's consider the next branch circuit (BC2).

BC2 (Row 2)

To Column 1 - From sub-panelboard to the first stub-up (Row 2, Column 1), the referenced PVC per Figure 49 will now contain 4 current-carrying conductors. The branch-circuit conductors routed to this stub-up must be sized to carry the load [17.3A] of 10 fixtures a distance of 192' [5' + 132' + 5' + 5' + 40' + 5'] at a voltage drop not exceeding 2 percent of the voltage at the sub-panelboard (203V x .02 = 4.06V) to realize the calculated minimum supply voltage (197.6V) at the farthest fixture pair of Row 2. Therefore, to determine the size branch-circuit conductors needed, the following methods are applied:

$$R = \frac{4.06V \times 1000'}{2 \times 192' \times 17.3A} = .6112\Omega \quad \text{(Use 6 AWG - [resistance .491}\Omega \text{ smaller than .6112}\Omega \text{])}$$

$$CM = \frac{2 \times 12.8\Omega \times 192ft \times 17.3A}{4.06V} = 20944 \; \textit{(rounded off)} \; \text{(reference use of 6 AWG)}$$

Based upon the selected use of 6 AWG conductors, because the conductors (two line-to-line) supplying the 10 fixtures will encounter an adverse ambient temperature while enclosed with two other current-carrying conductors (totaling four) both conditions must be factored in to determine whether these conductors are sufficient to carry the load. Per Tables 310.15(B)(2)(a) and 310.15(B)(3)(a), .94 and .80 are the factors that must be applied to determine the sufficiency of the conductors along with the provision for supplying a continuous load at 125 percent (1.25).

$$I = \frac{17.3A \times 1.25}{(.80 \times .94)} = 28.76A \; \text{(ampacity less than 6 AWG @ 75°C [65A], 6 AWG o.k.)}$$

Now that the sufficiency of the conductors has been verified, the voltage drop that the conductors will experience from the sub-panelboard to the first stub-up can be calculated,

$$V_D = \frac{2 \times .491\Omega \times 192' \times 17.3A}{1000'} = 3.26V \; \textit{(rounded off} \text{ - worst case to be applied)}$$

$$V_D = \frac{2 \times 12.8\Omega \times 192' \times 17.3A}{26240} = 3.24V \text{ (rounded off)}$$

which is used to determine the supply voltage of the branch-circuit conductors (BC2) at the first stub-up of Row 2.

$$203V - 3.26V = 199.74V$$

Now that the supply voltage at the *1st stub-up* has been determined, let's determine the supply voltage at the *last stub-up* (Column 5) using the same size 6 AWG branch-circuit conductors to determine whether the 5 percent limit between the feeder and branch circuit for this group of fixtures has been maintained. Again, from here on out only one method will be used to calculate the various voltage drops where such voltage drops are rounded off where needed.

(Stub-up at Column 2 - from Column 1)

$$V_D = \frac{2 \times .491\Omega \times {}^*50' \times 13.84A}{1000'} = .68V$$

$$199.74V - .68V = 199.06V \text{ (stub-up voltage)}$$

(Stub-up at Column 3 - from Column 2)

$$V_D = \frac{2 \times .491\Omega \times {}^*50' \times 10.38A}{1000'} = .51V$$

$$199.06V - .51V = 198.55V \text{ (stub-up voltage)}$$

(Stub-up at Column 4 - from Column 3)

$$V_D = \frac{2 \times .491\Omega \times {}^*50' \times 6.92A}{1000'} = .34V$$

$$198.55V - .34V = 198.21V \text{ (stub-up voltage)}$$

(Stub-up at Column 5 - from Column 4)

$$V_D = \frac{2 \times .491\Omega \times {}^*50' \times 3.46A}{1000'} = .17V$$

$$198.21V - .17V = 198.04V \text{ (stub-up voltage)}$$

*length of stub-ups and distance between poles

The difference between the initial supply voltage at 208V and 198.04V is 9.96V which results to an overall voltage drop of 4.79 percent (9.96V/208V). As for this branch circuit (BC2), 6 AWG conductors can be used to exceed the 5 percent limitation (197.6V).

Depending upon the desired voltage loss, the conductors extending from each stub-up to the paired fixtures can be sized smaller than 6 AWG.

Now let's consider the last branch circuit (BC3).

BC3 (Row 3)

To Column 1 - From sub-panelboard to the first stub-up (Row 3, Column 1), the referenced PVC per Figure 49 will now contain 2 current-carrying conductors. The branch-circuit conductors routed to this stub-up must be sized to carry the load [17.3A] of 10 fixtures a distance of 242' [5' + 132' + 5' +5' + 40' + 5' + 5'+ 40' + 5'] at a voltage drop not exceeding 2 percent of the voltage at the sub-panelboard (203V x .02 = 4.06V) to realize the calculated minimum supply voltage (197.6V) at the farthest fixture pair of Row 3. Therefore, to determine the size branch-circuit conductors needed, the following methods are applied:

$$\mathbf{R} = \frac{4.06V \times 1000'}{2 \times 242' \times 17.3A} = .4849\Omega \text{ (Use 4 AWG - [resistance .308}\Omega \text{ smaller than .4849}\Omega\text{])}$$

$$\mathbf{CM} = \frac{2 \times 12.8\Omega \times 242ft \times 17.3A}{4.06V} = 26398 \text{ (rounded off) (reference use of 4 AWG)}$$

Based upon the selected use of 4 AWG conductors, because the conductors (two line-to-line) supplying the 10 fixtures will encounter an adverse ambient temperature this condition must be factored in to determine whether these conductors are sufficient to carry the load. Per Table 310.15(B)(2)(a), .94 is now the only factor that must be applied to determine the sufficiency of the conductors along with the provision for supplying a continuous load at 125 percent (1.25).

$$\mathbf{I} = \frac{17.3A \times 1.25}{.94} = 23A \text{ (ampacity less than 4 AWG @ 75°C [85A], 4 AWG o.k.)}$$

Now that the sufficiency of the conductors has been verified, the voltage drop that the conductors will experience from the sub-panelboard to the first stub-up can be calculated,

$$\mathbf{V_D} = \frac{2 \times .308\Omega \times 242' \times 17.3A}{1000'} = 2.58V \text{ (rounded off - worst case to be applied)}$$

$$\mathbf{V_D} = \frac{2 \times 12.8\Omega \times 242' \times 17.3A}{41740} = 2.57V \text{ (rounded off)}$$

which is used to determine the supply voltage of the branch-circuit conductors (BC3) at the first stub-up of Row 3.

$$203V - 2.58V = 200.42V$$

Now that the supply voltage at the *1st stub-up* has been determined, let's determine the supply voltage at the *last stub-up* (Column 5) using the same size 4 AWG branch-circuit conductors to determine whether the 5 percent limit between the feeder and branch circuit for this group of fixtures has been maintained.

(Stub-up at Column 2 - from Column 1)

$$V_D = \frac{2 \times .308\Omega \times *50' \times 13.84A}{1000'} = .43V$$

200.42V − .43V = 199.99V (stub-up voltage)

(Stub-up at Column 3 - from Column 2)

$$V_D = \frac{2 \times .308\Omega \times *50' \times 10.38A}{1000'} = .32V$$

199.99V − .32V = 199.67V (stub-up voltage)

(Stub-up at Column 4 - from Column 3)

$$V_D = \frac{2 \times .308\Omega \times *50' \times 6.92A}{1000'} = .21V$$

199.67V − .21V = 199.46V (stub-up voltage)

(Stub-up at Column 5 - from Column 4)

$$V_D = \frac{2 \times .308\Omega \times *50' \times 3.46A}{1000'} = .11V$$

199.46V − .11V = 199.35V (stub-up voltage)

*length of stub-ups and distance between poles

The difference between the initial supply voltage at 208V and 199.35V is 8.65V which results to an overall voltage drop of 4.16 percent (8.65V /208V). As for this branch circuit (BC3), 4 AWG conductors can be used to exceed the 5 percent limitation (197.6V).

Depending upon the desired voltage loss, the conductors extending from each stub-up to the paired fixtures can be sized smaller than 4 AWG.

ABOUT THE AUTHOR

For over thirteen years, Alvin Walker, a native of Shreveport, Louisiana owned and operated a small yet successful electrical contracting business. He now works as an author and instructor specializing in electrical and NEC training where his services are available throughout the United States. In his over thirty year of experience, he has developed a very strong background in electrical engineering, electrical design, electrical maintenance and construction. He has taught Business Law for Contractors, the National Electrical Code, Electrical Theory, and other basic and advanced electrical classes at Bossier Parish Community College, Louisiana State University-Shreveport, Northwest Louisiana Technical College (Forcht Wade Correction Center), and Southern University-Shreveport to include once serving as the Department Head of Industrial Electricity at Houston Community College-Stafford, Texas.

Mr. Walker is best known for his hands-on approach and the ability to simplify and explain the most difficult electrical subject matters. He is a master electrician and holds a Louisiana state license as an electrical contractor. He has a degree in electrical engineering from the University of South of Carolina and has worked as an electrical engineer for E.I. DuPont and Westinghouse at The Savannah River Plant (Company) of Aiken, South Carolina and M.W. Kellogg of Houston, Texas.

In his daily life Mr. Walker is a devoted Christian who has a passion for serving Christ, his fellowman and teaching and spreading the Word of God. As a recipient of three honorable discharges, he served over 9 years in the United States Army.

He enjoys traveling, wood-works and carpentry but is best known for his famous smoked barbeque ribs and sweet ice tea.

www.ingramcontent.com/pod-product-compliance
Lightning Source LLC
Chambersburg PA
CBHW042031220326
41598CB00073BA/7449